What Did Dinosaurs Think About?

ANIMAL WORLDS

Jessica Serra, Series Editor

*What if,
instead of looking at animals
through our own eyes,
we looked through theirs?*

Recent scientific discoveries offer us a new perspective on the animal kingdom, shattering the myth that once equated animal behavior with that of machines. We now know that humans are not the only beings with intelligence, emotions, and language skills.

Even though animals share our environment, they perceive and understand it in their own way. Equipped with specific sensory equipment, they selectively pick up certain meaningful signals and evolve in a world of their own. This means that our human world is only one among millions of others.

Shifting our perspective to reflect this reality forces us to rethink our own place in the world, not as superior to other living beings but amid them. This perspective also allows us to discover the infinite richness of animal lives and the dazzling complexity of "beasts."

Enlightened by science, this series endeavors to open doors to these other worlds by providing a new understanding of living things and, therefore, a new understanding of ourselves.

Also in the Series
Jessica Serra, *The Beast Within: Humans as Animals* (2024)
Mathieu Lihoreau, *What Do Bees Think About?* (2024)

What Did Dinosaurs Think About?

Jean Le Loeuff

Translated by Alison Duncan

Johns Hopkins University Press
BALTIMORE

© 2025 Johns Hopkins University Press
All rights reserved. Published 2025
Printed in the United States of America on acid-free paper
9 8 7 6 5 4 3 2 1

This work was originally published in French as *Dans la peau d'un dinosaure*
© humenSciences/Humensis, 2023.

Johns Hopkins University Press
2715 North Charles Street
Baltimore, Maryland 21218
www.press.jhu.edu

Library of Congress Cataloging-in-Publication Data

Names: Le Lœuff, Jean, author. | Duncan, Alison, translator.
Title: What did dinosaurs think about? / Jean Le Loeuff ; translated by Alison Duncan.
Description: Baltimore : Johns Hopkins University Press, 2025. | Series: Animal worlds | Includes bibliographical references. | Identifiers: LCCN 2024055683 (print) | LCCN 2024055684 (ebook) | ISBN 9781421452074 (paperback) | ISBN 9781421452081 (ebook)
Subjects: LCSH: Dinosaurs—Psychology. | Animal intelligence. | Geology, Stratigraphic.
Classification: LCC QE861.6.B44 L45 2025 (print) | LCC QE861.6.B44 (ebook) | DDC 567.9—dc23/eng/20250529
LC record available at https://lccn.loc.gov/2024055683
LC ebook record available at https://lccn.loc.gov/2024055684

Special discounts are available for bulk purchases of this book.
For more information, please contact Special Sales at specialsales@jh.edu.

EU GPSR Authorized Representative
ÐOGOS EUROPE, 9 rue Nicolas Poussin
17000, La Rochelle, France
e-mail: Contact@logoseurope.eu

CONTENTS

SERIES EDITOR'S FOREWORD VII

Prologue 1
THE LIFE OF DINOSAURS, FROM CINEMA TO SCIENCE 1

1 Dinosauria: Preliminary Concepts 6
DINOSAURS' PLACE ON THE VERTEBRATE FAMILY TREE 9

2 Dinosaur Meninges 14
IGUANODON AT THE BEACH 16
AND ONE, AND TWO, AND THREE BRAINS 19
MEASURING INTELLIGENCE 25
ENCEPHALIZATION QUOTIENT AND ITS VARIANTS 27
THE SIZE OF A WALNUT 35

3 In Search of Lost Senses 36
THE SCENT OF HADROSAURS IN THE AIR 37
SEE *T. REX* AND DIE 43
THE SOPRANO AND THE RAPTOR 48
THE SONG OF THE DINOSAURS 52
AMPELOSAURUS COULDN'T SAY NO 55
TYRANNOSAUR KISSES 56
DINOSAUR NOCICEPTION 57
ARGENTINOSAURUS'S SIESTA 61

4 Mesozoic Sociology 65
ANTISOCIAL DINOSAURS 66
LIFELONG GROUPS 68
THE CRETACEOUS WILDEBEEST 72
YOUTH GANGS 74
RECOGNITION SIGNALS 79

CONTENTS

HUNTING   81
ABSENCE OF PROOF AND PROOF OF ABSENCE   84
DEFENSE STRATEGIES   85
PARASAUROLOPHUS'S DIETARY SUPPLEMENTS   94
TONGUE-TIED   96

5 Banter Between Lovers   98
COURTSHIP RITUALS   99
THE AGE OF CONSENT   103
MATING   105
THE INTERIOR OF A DINOSAUR'S CLOACA   106
AFTER LOVE   111
HEYUANNIA'S BLUE EGGS   116
GOOD MOTHERS   119
LIFE IN THE EGG   120
DID *T. REX* PLAY? AN OUTLANDISH HYPOTHESIS   123
PAINFUL LESSONS   125

Epilogue   126
FINAL THOUGHTS   126

Appendix   129
THE DINOSAUR FAMILY TREE   129
  THE VERY FIRST DINOSAUR   129
  THEROPODS   132
  SAUROPODOMORPHS   138
  THYREOPHORANS   141
  ORNITHOPODS   143
  MARGINOCEPHALIANS   144

ACKNOWLEDGMENTS   145
NOTES   147

SERIES EDITOR'S FOREWORD

Like many children, I was fascinated by dinosaurs from an early age. Where did this passion come from? Was it influenced by a fad in literature and film? My interest in the animal kingdom? My affection for lost worlds? Or perhaps it was the magnetic pull of these prehistoric monsters themselves? Undoubtedly, it was for all these reasons that I immersed myself in books on dinosaurs for hours on end.

My first dinosaur encounters aroused a thousand emotions, from amazement to shivers of fear. Shrouded in mystery, these creatures seemed unreal, blending with the make-believe monsters that scared me. More than at any other stage of life, a child thirsts for the imaginary, sometimes even intertwining what is consciously perceived and imagined.* Unlike adults, a child's forays into this realm don't suffer from a psychological barrier, as if a young brain needs to consume chimeras to grow and form. The world of dinosaurs is full of peculiar creatures with complex physiques. Their fantastical morphology and that they belong to vanished worlds make them precious imaginary material. Beyond the mythical, dinosaur symbolism has a powerful and ferocious quality. Next to a *T. rex*, a human being appears frail and helpless. Terrifying domination! And it's not just that. Although they reigned supreme for millions of

---

* For Jean-Paul Sartre, the imaginary is an intentional consciousness. While poorer in essence than detail-rich perceptive consciousness, intentional consciousness is creative and enables us to entangle different chimerical images. Children have the distinctive ability to introduce elements from their imaginary world into reality (for example, by inventing a fictitious friend they talk to for hours on end).

SERIES EDITOR'S FOREWORD

years, dinosaurs also epitomize *another* time, a time *before* humanity. It's now time to set the record straight. Even their extinction raises questions! Despite their strength, despite their evolutionary success, with a few exceptions, these prehistoric colossi all died from a meteorite impact. Like a mirror, their existence and then extinction remind us of our own insignificance and perishability, two things that are unbearable for humankind.

Many try to rationalize this fear by convincing themselves that, whether by comparing humans to prehistoric or modern-day animals, the human essence is different. For most of the 20th century, dinosaurs were thought to have had the same intelligence as contemporary reptiles—close to zero! Fortunately for us, paleontologist Jean Le Loeuff pays no mind to preconceived notions. With the precision of a watchmaker, he has scanned dozens of dinosaur skulls to recreate million-year-old brains in three dimensions. In this book he deciphers how each cerebral area functioned, while poking fun at some of the earlier hasty interpretations and presenting us with a new and unique portrait of dinosaurs. In his charming style, Le Loeuff takes us on an incredible journey through time as he tackles the ambitious challenge of reconstructing the behavior of animals that lived more than sixty-six million years ago.

Based on paleoneurological and paleogenetic discoveries, Le Loeuff transcends the descriptive study of fossils and lays the foundations for a new discipline: paleoethology. It's not an easy task. Contrary to "traditional" ethology, there aren't any firsthand witnesses to dinosaur behavior. Without ever compromising scientific rigor or glossing over hypotheses, Le Loeuff rises to the challenge and brings these fabulous creatures back to life. Opening the doors to a distant past, he awakens our curiosity as much as he delights our childlike sense of wonder.

Were these lost worlds populated only by unintelligent beasts? The life of a dinosaur was far from easy. Le Loeuff recounts the incredible discovery of hadrosaur skulls that survived multiple traumatic injuries, suggesting "the possibility of mutual aid during convalescence." So "dinosaur" isn't synonymous with "survival of the fittest"! As Peter Kropotkin asserted, the emergence of cooperation played an important, if not a more important, role in evolution than competition. The existence of family groups, packs, and "youth gangs" made it highly likely that the group would be protected from predator attacks, and that group members would cooperate for hunting or raising young. Far from the dim-witted reputation that clings to them, most dinosaurs undoubtedly experienced emotions, formed bonds, and exceeded our expectations in terms of cognitive abilities. More importantly, these worlds uninhabited by humans were not devoid of humanity,* in the virtuous sense of the word. That's the paradox we must confront. Thinking about the world as it *was* means thinking about a universe that didn't need human beings to exist. Whether we existed or not, the Earth was turning. It was home to intelligent life forms for such a long time before us that the history of humankind pales in comparison! In this way of looking at time, *Homo sapiens* doesn't mark the beginning or the end. It's simply one branch among many, embedded in an evolutionary continuum that isn't static. Paleontology not only revives memories erased by death but also invites us to think about the order of the world. Which is unsettling, given a lack of any "final determination."[1]

But rather than lose ourselves in the absurd, couldn't we find meaning instead? Since in humans "evolution becomes aware of itself,"† how can we not be moved by the incalculable number of

---

* While altruism and empathy exist in many modern and prehistoric animals, here the word "humanity" has been deliberately chosen to deconstruct the systematic association we make between human beings and the capacity to be benevolent and to feel compassion toward others.
† Biochemist Hans Tuppy coined the phrase.

events it took for us to exist today alongside chimpanzees, elephants, and the descendants of dinosaurs? Or not be moved by our collective fragility and our interdependence? For the first time since life began on Earth, the human species has the extraordinary power to bend the curve of its destiny. It remains to be seen what we'll do with it.

Jessica Serra

What Did Dinosaurs Think About?

PROLOGUE

The Life of Dinosaurs, from Cinema to Science

If most people on Earth believe that velociraptors were pack hunters, it's thanks to Steven Spielberg and *Jurassic Park*. In the film, not only were the terrifying raptors able to use their clawed hands to open doors, but they also excelled at setting deadly traps for their pursuers. They killed the park's experienced game warden before he could even put up a fight—and they did it by using extraordinarily complex strategies. Strategies that are, frankly, unknown to present-day predators, whether reptiles or mammals, with one exception: *Homo sapiens*. In each subsequent film in the series, the raptors seem to become increasingly clever and dangerous.

Are these characteristics the poetic inventions of Michael Crichton (the author of the *Jurassic Park* novel on which the film is based) or scientific certainties? My goal for this book is to answer this reasonable question by examining the most recent discoveries in paleontology. Scientists have historically looked down on such paleobehavioral and paleoethological questions, but now in the third decade of the 21st century these issues have become central in paleontological research. This book therefore examines dinosaurs through a unique lens. What did they smell, hear, and see? What kinds of sounds did they make? Did they live alone or in groups?

Did they feel pain? These and so many other questions have long been more within the purview of novelists and filmmakers than of paleontologists.

The study of the behavior of present-day animals (ethology) is still in its infancy. Research on animal behavior is based on the observation of wild animal populations, whether actually in the wild or in captivity, but it's also interested in domestic animals. Recently we've begun to better understand the behavior of some mammals—like cats, rats, and dogs—and birds, such as corvids (the crow family) and psittacines (the parrot family). In other words, we tend to study animals we find rather cute. But the study of animals we typically find repulsive, like reptiles, is even more nascent, although what we have learned has already challenged a number of preconceived ideas. Not so long ago, reptiles were considered to be slow, unintelligent, solitary machine-like beings, but in recent years, ethologists have discovered them to be social creatures, capable of learning and behaving in quite complex ways. Paradoxically, even though dinosaurs are reptiles, they have become one of the most popular children's toys in recent years, surpassing teddy bears and other cuddly animals. Is there a child's room that doesn't have a collection of *T. rex* figures, a triceratops stuffed animal, or a dinosaur encyclopedia? Fortunately not! Now that dinosaurs are part of our (collective) imagination starting at a young age, it makes sense that we want to know what was going on in their brains, just as we want to understand everything about the behavior of cute kittens. "People want to know!" my editor tells me. Well then, that's what this book will endeavor to address.

We'll give it a try at least, but seeing as we're still feeling around in the dark trying to understand the behavior of animals right in front of us, what can we expect to learn about the behavior of creatures that disappeared 66 million years ago and that we can no longer observe evolving, reproducing, making noises, playing—or more succinctly, living? Fortunately for me, paleoethology can take a few indirect paths, which this book proposes to explore through two major routes. The first is by studying dinosaurs' closest living relatives: their crocodilian cousins (whose common ancestor with dinosaurs lived around 245 million years ago) and the eleven thousand species of birds who are the living representatives of the dinosaur group. The latter are the only group of dinosaurs to have survived the ecological disaster that wiped out their cousins at the end of the Cretaceous Period. But remember, while all birds are dinosaurs, the opposite is not true. Most dinosaurs were non-avian, and these are the ones we'll be focusing on here: *Triceratops*, *Parasaurolophus*, *Stegosaurus*, *Tyrannosaurus*, and *Diplodocus*. The second route is based on the in-depth study of the fossil remains accumulated by paleontologists over the past two centuries. The new interest in paleobehavior is largely due to the new imaging techniques that have recently become available, particularly X-ray microtomography (or micro-CT), which allow us to access the inside of fossilized skulls and study the shape of the brain and inner ear of different groups of dinosaurs. This ongoing technological revolution has allowed us to formulate a plethora of new hypotheses that were inconceivable at the end of the 20th century. We now also know the color of several dinosaur species thanks to scanning electron microscopy, and the color of their eggs owing to Raman microspectroscopy. We can conjure their sensory capabilities and soon even their vocalizations.

Alongside these technological achievements, traditional paleontological methods are yielding a wealth of new information about dinosaur behavior. Paleoichnology (the study of fossilized footprints) allows us to examine different aspects of their sociality. Both group living and social distancing seem to have been strategies they used. Analyses of predation marks on skeletons, combined with work on coprolites (fossilized feces), are as much a window on dinosaurs' sometimes highly selective meals as on their hunting techniques. The discovery over the last quarter century of countless egg-laying sites containing nests, eggs, and fossilized fetuses has also provided a wealth of new information on reproduction, nest surveillance strategies, parental care, egg incubation periods, and even in ovo communication (while fetuses were still in the egg, that is).

In short, we're moving away from the actualistic approach of fumbling around in the dark (in other words, inferring dinosaur behavior solely from that of crocodilians) toward directly understanding the way of life of the countless dinosaur species that populated Earth for 170 million years. It's a matter of fumbling around in the dark because, when it comes to actualism, to which group of animals should we liken dinosaurs? For most of the 20th century, dinosaurs were compared to crocodilians or even lizards (which are much more distant cousins), meaning they were saddled with the supposed defects of present-day reptiles. Those were the days when the poor sauropods were depicted as living in lakes, using their long necks as snorkels. There was a time, thankfully long gone, when it was believed that the same sauropods were devoured starting at the end of their tails long before that information reached their brains and that stegosaurs had a second brain in their pelvis. We'll dispense with all these urban legends here. "Dinosaurs . . . were unintelligent beings," said French paleontologist Marcellin Boule in 1905, summing up the long-held scant regard for our protagonists' intellectual performance. Then, during the "dinosaur renaissance" of the 1970s, these animals were suddenly compared to mammals and birds in terms of their intellectual and physical capacities, becoming fast, lively, intelligent beings with robust social interactions. *Jurassic World*'s raptors are, therefore, downright cleverer than the military, which may be taking things a little too far.

What's the real story? Were dinosaurs unintelligent or sharp as tacks? What we currently know about the neuroanatomy of some

PROLOGUE: THE LIFE OF DINOSAURS, FROM CINEMA TO SCIENCE

dinosaur species may suggest a more nuanced reality. Non-avian dinosaurs were neither mammals nor birds, but their brains were also very different from those of crocodilians, whose behavior recent ethological research has also shown to be far more complex than previously imagined.

What was going on in the minds of these extinct animals? This book provides an overview of what we know and don't know in answer to this fascinating question. Paleontologists (or paleoethologists, to be more precise) seeking to understand the behavior of extinct rhinoceroses or felines can reasonably base their work on the ethology of present-day representatives of these mammalian groups. But how does that help us infer the behavior of *Triceratops*, for example, the fascinating three-horned herbivore that weighed over ten metric tons? Neither crocodilians nor birds are suitable present-day equivalents. Only by synthesizing paleoneurological and paleontological data can we gain a clearer picture.

Many of the current hypotheses concerning dinosaur behavior are presented here. These are not ill-defined conjectures or gratuitous speculations but scientific hypotheses based on facts and observations, some of which will inevitably be disproved in the years to come, while others will be reinforced.

As already mentioned, dinosaur paleoneurology is rapidly expanding. New scientific papers are published every month, describing newly discovered braincases (or crania), the part of the skull that once enclosed the brain, in ever greater detail. There's no question that new hypotheses about the behavior of these "giant lizards" will soon abound!

1

Dinosauria

Preliminary Concepts

Before we delve into dinosaurs' intellectual prowess or their social and mating behavior, it's worth taking a quick look at what dinosaurs were in the first place. The current definition of the zoological group Dinosauria describes dinosaurs as the most recent common ancestor of *Triceratops horridus* (a Cretaceous horned animal) and *Passer domesticus* (the sparrow), as well as all of this common ancestor's descendants. It's as simple as that. This definition supports a considerable number of variations, as long as you choose the most recent common ancestor of an ornithischian (*Triceratops, Iguanodon, Stegosaurus*, etc.) and a saurischian (tyrannosaur, sparrow, chicken, *Diplodocus*, or pigeon—they're all part of the same classification). Dinosaurs represent a branch of the vertebrate phylogenetic tree that has enjoyed considerable success over the past 235 million years. In fact, there are just over eleven thousand species of dinosaurs still living today—birds. As ornithology is not the subject of this book, we'll be discussing non-avian dinosaurs—that is, all dinosaurs except birds. The last common ancestor of *Triceratops* and *Diplodocus* probably lived 240 million years ago, and these different groups of dinosaurs evolved separately over tens of millions of years, inevitably leading to considerable variations in behavior. If you need convincing, just remember that humans share a common ancestor with whales, mammoths, and rabbits that lived around 65 million years ago.[1]

All the events discussed here took place during the Mesozoic Era (sometimes still referred to as the Secondary Era), which occurred 251.9 to 66 million years ago. This geologic era began after the mass extinction at the end of the Paleozoic Era, when 90% of species disappeared. The Mesozoic Era is divided into three periods: Triassic (251–201 Ma), Jurassic (201–145 Ma), and Cretaceous (145–66 Ma). These periods are in turn subdivided into epochs (Early Triassic, Middle Triassic, Late Triassic, etc.), and epochs into ages. Each age corresponds to a geologic stage lasting an average of 5 to 6 million years. The Mesozoic Era ended with another major mass extinction at the Cretaceous-Paleogene (or Cretaceous-Tertiary) boundary 66 million years ago. The impact of a huge meteorite caused fatal consequences for dinosaurs, resulting in the definitive extinction of all non-avian dinosaurs. Consequently, this book won't be looking at the events of the Cenozoic Era, which is still unfolding as you read these lines. The current period, the Quaternary, is just one period of this era.

By occupying the planet for 170 million years, dinosaurs unwittingly witnessed major geological events, including the most recent continental drift. Bear in mind that the Atlantic Ocean is broadening at a rate of 2 centimeters (approximately three-quarters of an inch) per year, which means today you only need one more oar stroke to cross the ocean than your ancestor five generations ago did (or try flying, it's faster). But 2 centimeters per year corresponds to 2 million centimeters per million years, or 20 kilometers (almost 12.5 miles). And in 170 million years, that would be 3,400 kilometers (over 2,100 miles), which is starting to add up to quite a lot of oar strokes. When the first dinosaurs appeared in the Triassic Period, they unknowingly found themselves on a single gigantic continent, called Pangaea, which grouped all the continental masses into a single supercontinent. This mass began to break up as early as the Jurassic Period with the opening of the Central Atlantic and the Indian Ocean, and then the South Atlantic during the Cretaceous Period, and so on. In short, the different continental masses we know today gradually separated from each other over the course of the Mesozoic Era. Each continental mass took its own dinosaur population with it, which then evolved independently, resulting in endemism during the Cretaceous Period and meaning that very different dinosaurs were found on the various landmasses.

The data we have on dinosaur behavior comes first and foremost from the countless fossils discovered over the last two centuries, most of which have been carefully preserved in the collections of hundreds of museums and universities around the world. Other fossils are in private collections, and these don't contribute to scientific research since paleontologists can't study them. Only fossils from collections that are accessible to researchers and whose continued accessibility is ensured can be published in scientific journals. It's hardly surprising then that paleontologists are less than enthusiastic about fossil auctions—so many specimens are lost to science for good. Among the properly preserved remains are several million bones and a few thousand skeletons of varying completeness, as well as numerous nests and eggs and a plethora of fossilized footprints. Much rarer are coprolites (fossilized dinosaur feces), regurgitalites (fossilized vomit—yes, it exists!), and gastroliths (stones ingested by certain dinosaurs and found in their stomachs).[2] Skin prints are also rare. They are invaluable for learning about dinosaur integument (scales or feathers) and, therefore, dinosaurs' color. By 2022, around thirteen hundred dinosaur species had been identified by paleontologists, with almost fifty new species added to the list every year. These thirteen hundred species represent only a tiny fraction of the dinosaurs that actually populated

DINOSAURIA: PRELIMINARY CONCEPTS

the planet for 170 million years; tens of thousands remain undiscovered or didn't leave fossils. Dinosaurs have been discovered in rocks dating back 235 to 66 million years on every continent and in about sixty countries around the world. This considerable amount of data, patiently tinkered with by generations of paleontologists, has led to a host of hypotheses on subjects as bizarre as *T. rex*'s playfulness, titanosaurs' social interactions, and *Parasaurolophus*'s* dietary supplements. This data has also led to a better understanding of the dinosaur family tree, a vast undertaking that has been steadily refined over the past 150 years, especially since the advent of phylogenetic classification in the 1990s.

Dinosaurs' Place on the Vertebrate Family Tree

To understand the kinship between dinosaurs and present-day animals, we need to take a look at the phylogeny of amniotes. Amniotes are vertebrates whose young develop in amniotic fluid. This group excludes amphibians and, of course, "fish." Today's amniotes include mammals (6,500 species, half of which are bats and rodents), birds (11,150 species), and "reptiles" (10,600 species). This may be a hit to our mammal-centric ego, but clearly our lineage is by no means the dominant one in the amniote world!

Why put quotation marks around "reptiles" and "fish"? Because they are not monophyletic groups, a concept that requires some explanation. As phylogenetic inference will be discussed at length in the following lines, it's essential to keep in mind a brief definition of the principle: zoological groups must reflect the family tree of

---

* *Parasaurolophus* is a herbivore from the Late Cretaceous Period (73 to 77 million years ago). This hadrosaur had a remarkable curved crest measuring over 2 meters (over 6.5 feet) long on the top of its skull.

living organisms, and a group must therefore include the most recent common ancestor and all the descendants of this common ancestor. This is called a "monophyletic group." This of course challenges the old Linnaean classification system, which included classes of Reptiles and Birds, for example. The common ancestor of turtles, lizards, and crocodilians, however, is shared with birds, and therefore the present-day reptile group includes birds. The traditional usage of the term "reptile" (which excludes birds) is no longer scientifically correct, since bygone "reptiles" are a paraphyletic group—meaning, the group includes some but not all of the common ancestor's descendants. As we'll see later, most researchers currently use the term "sauropsid," which includes "reptiles" and birds. Note that the same verdict applies to "fish," since the descendants of the common ancestor of all present-day "fish" include tetrapods (amphibians, sauropsids, and mammals). Let's put the drama into perspective: there's no need for your local fish market to add quotation marks to its sign; only the most rigid of phylogeneticists would jump down your throat for saying you eat fish. If, however, you want to avoid upsetting people with odd habits, be aware that omitting the quotation marks when talking about fish can cause serious offense (all jokes aside). The precise thing to say is that you eat actinopterygians, a monophyletic group that contains almost all edible fish (except sharks and rays, which are chondrichthyans) and excludes tetrapods. Clearly a phylogenetically correct fish market would need to be renamed "actinopterygian market." Mammals, on the other hand, don't need quotation marks as they constitute a monophyletic group that includes the common ancestor of monotremes (platypuses), marsupials (koalas), and placentals (humans), and all of its descendants.

Let's take a moment to look at the evolution of these major vertebrate lineages in the Late Paleozoic Era, when amniotes appeared and diversified. As amniotes evolved, beginning around 315 million years ago at the end of the Carboniferous Period, they rapidly diverged into two major groups: synapsids, represented today by mammals, and sauropsids, represented today by turtles, lizards, snakes, crocodilians, and birds. At that time, they were all just insignificant little creatures vaguely resembling small lizards. The sauropsid lineage, with its 21,750 present-day species, is the one we're interested in. To simplify, it's divided into lepidosauromorphs (present-day lizards, snakes, and amphisbaenians,* and also tuatara† endemic to New Zealand) and archosauromorphs (including crocodilians, birds, dinosaurs, and perhaps turtles). These two major sauropsid groups diverged between 259 and 285 million years ago, a few million years before the end of the Paleozoic Era.[3] One of the subgroups of archosauromorphs are archosaurs, which encompasses living crocodilians and birds, along with dinosaurs and other extinct groups such as pterosaurs (flying reptiles). The oldest known archosaur dates from the Olenekian Age, the second age in the Early Triassic Epoch (251–247 Ma). The only non-dinosaurian archosaurs alive today are crocodilians, which are the closest living cousins of dinosaurs (birds, on the other hand, are dinosaurian archosaurs). Other archosauromorphs still exist and are the survivors of another large subgroup, called the Pantestudines. This group includes modern turtles, which most recent studies suggest are the closest living cousins of crocodilians. Some current phylogenetic inferences therefore place turtles within the Pantestudine group along with a large percentage of Mesozoic marine reptiles (particularly plesiosaurs and ichthyosaurs). Other inferences place turtles in a separate

---

* Close cousins of snakes, amphisbaenians are vermiform burrowing animals found in warm regions. Because they spend their lives underground, they are rarely encountered.

† Tuatara (or *Sphenodon*) are the only living rhynchocephalians, the last remnant of a group that was highly diverse during the Mesozoic Era. While tuatara resemble lizards, they're not lizards, just distant great-uncles.

group from lepidosaurs and archosauromorphs, while others still place marine reptiles (sauropterygians) in a separate sauropsid lineage. In short, there's still a lot of uncertainty out there—and don't worry, that's perfectly normal! It's complicated enough to establish the phylogeny of living animals whose complete anatomy and protein and DNA sequences we can analyze; so when all we have are bits of skeletons, let's just say we have to remain humble about the strength of our phylogenetic inferences.

So, from a phylogenetic point of view, when we're interested in dinosaur behavior, it makes sense to look at crocodilian behavior. At any rate, it makes better sense than looking at turtle behavior and even better sense than looking at lizard behavior, since a lizard is a more distant relative. Because birds are theropod dinosaurs and are the only living dinosaurs, it's equally tempting to look at their lifestyles and infer those of their distant relatives. But we must bear in mind that modern birds diverged a very long time ago—during the Jurassic Period—and their lifestyle led to considerable physiological changes. It's probably no more reasonable to extrapolate the observations made on the wren to a 10-metric-ton triceratops than it is to think the behavior of primates is the key to that of rabbits (which, as you may have guessed, are our close relatives within the class Mammalia, and more specifically in the Euarchontoglire group, because you and I, and rabbits too, are Euarchontoglires according to mammalogists).[4] This reservation aside, actualist dinosaur comparisons will be preferentially chosen within these two sauropsid groups: crocodilians and birds.

All dinosaur specialists agree on classifying them into five major clades, each of which evolved from a single ancestor. These five groups, of which I'll mention a few emblematic representatives in parentheses, are theropods (*Tyrannosaurus, Velociraptor, Gallus*), sauropodomorphs (*Diplodocus, Brachiosaurus*), thyreophorans (*Stegosaurus, Ankylosaurus*), ornithopods (*Iguanodon, Parasau-*

rolophus), and marginocephalians (*Triceratops, Pachycephalosaurus, Psittacosaurus*). According to most specialists, theropods and sauropodomorphs have a common origin within saurischians, while the other three groups make up ornithischians. Ornithopods and marginocephalians have a more recent common ancestor, however, than the one they share with thyreophorans. These different branches evolved independently of each other for 130 to 170 million years, so by the end of the Mesozoic Era, their representatives had been evolutionarily and genetically separated for much longer than our primate group has been from whales or rabbits. As a result, the behavioral inferences that can reasonably be proposed on the basis of available data for a dinosaur like *Diplodocus* obviously don't apply to *Triceratops*, and vice versa. Talking about "dinosaur" behavior is as absurd as talking about "mammal" behavior. In the following chapters, we'll be making deductions about the senses of smell, hearing, and sight of specific species that don't apply to the entire zoological group.

2

Dinosaur Meninges

In recent decades, many paleontologists have concentrated their efforts on classifying dinosaurs (see the appendix, The Dinosaur Family Tree, for an overview), with results that are still rather tenuous. The trend in recent years, however, has shifted to understanding dinosaur paleobiology. Without being able to directly observe their lifestyles—our subjects being extinct makes this definitively unattainable—a careful study of their fossilized remains fortunately enables us to formulate some hypotheses. And the first organ of interest in our quest is, of course, the seat of all activity: the brain. Like the lungs, liver, or heart, the brain is made up of soft tissue, which decomposes first after an animal dies. No dinosaur liver, heart, or brain has yet been found in fossil form, or at least none has been found yet with any certainty. (In this field, definitive conclusions are imprudent, but it is possible a small piece of brain has been found—more on that later.) Unlike other organs, the brain is enclosed in a bony case, which does commonly fossilize. By cutting the skull in half from the back, the cranial cavity can be found, and by removing the rock matrix filling it, the cavity's bony walls and many openings can be uncovered. During the dinosaur's lifetime, nerves and blood vessels passed through these structures. A cast of the now-empty cranial cavity can be made, which will approximate the brain's shape. It's only approximate because the brain of a living animal is surrounded by both the meninges, which can be relatively thick, and numerous blood vessels. An endocranial cast can therefore be likened to a pie, with the pie filling representing the gray matter and the crust representing the meninges. The thicker the crust, the less detail will be obtained about the filling inside.

In recent years, a far more effective method than aggressive skull sawing has been developed thanks to the invention of CT scans and 3D visualization tools at the turn of the 21st century. These tools have made it possible to study a skull's internal structures without having to damage the skull. This technological revolution gave paleoneurology a sudden boost and, incidentally, made it possible to write this book. These days, it's commonplace to refer to endocranial scans in the description of a dinosaur. Dinosaur paleoneurology is trendy! "Digital dissection" is a new anatomical tool, and according to Lawrence Witmer, a pioneer in the discipline, this invention is just as important as that of the scalpel in terms of scientific impact.[1] And it's true this technique has led to some amazing advances in understanding dinosaur anatomy.

An endocranial cast, whether obtained by the traditional method or by a scan, is an approximation of the brain itself and is detailed enough to distinguish a few major brain areas. Accustomed as we are to seeing lambs' brains at the butcher's or pictures of the human brain, dinosaur brain anatomy is not necessarily familiar looking to us. Our mammalian brains consist of two large connected hemispheres, with the other brain components hidden beneath this large mass. While the same part of the brain (the hemispheres) is also highly developed in birds, that isn't the case in other archosaurs. Since butchers don't sell crocodile brains, we're less accustomed to their much more complex appearance and much smaller size. The brain occupies only a small part of the back of the skull in both crocodilians and dinosaurs. When dissecting a crocodilian brain, from front to back we can distinguish the olfactory bulbs (lobes) connected by long olfactory tracts to very small cerebral hemispheres, a pituitary gland* below, and finally the optic lobes and cerebellum. When the cranial cavity of the same crocodilian is scanned, however, the picture becomes much less clear: the olfactory lobes and hemispheres can still be distinguished, as can the pituitary fossa, but the posterior areas of the brain (optic lobes and cerebellum) are no longer discernible owing to the thickness of the tissue around the brain itself. And that's pretty much what we find in most dinosaurs. Depending on the degree of development of these different areas, it's possible to deduce certain aspects of a dinosaur's biology, such as its senses of smell, sight, and hearing. The study of the senses in different dinosaur species will be addressed in the next chapter; here we'll focus on dinosaur brain size and how it relates to their intelligence.

Iguanodon at the Beach

Dinosaur paleoneurology has long been a fringe area of study because of a lack of material and adequate means to analyze it. You either had to be lucky enough to come across a natural endocranial cast or a broken skull, or have the audacity to cut open a skull to see what was inside. As you might imagine, this last option would make the average museum curator shriek in horror. An even more appalling method called "serial sectioning" was widely used in the 20th century by Swedish paleontologists specialized in Paleozoic fish. The technique involves polishing a fossil by removing sections measuring a few microns or millimeters in sequence, and then

---

* The pituitary gland (or hypophysis) is located at the base of the vertebrate brain and secretes numerous hormones.

drawing each cut section. These drawings create a 3D reconstruction of the fossil's internal anatomy. Serial sectioning has led to huge advances in paleoichthyology (paleontology of "fish"). Of course in the end, skulls studied this way become a pile of dust. To my knowledge, no dinosaur skull has ever been subjected to this atrocious method, but a few have definitely been sawed in half to access the cranial cavity.

In order to start studying the dinosaur brain, it was necessary to find suitable dinosaur skulls, and in particular the back part of skulls, which takes us to Victorian England in the 1860s. The first dinosaur discoveries had taken place on these very lands forty years earlier. They consisted of a number of vertebrae and limb bones, with only a few fragments of *Megalosaurus* and *Iguanodon* jawbones, and therefore no potential brains at all.

One of the first dinosaur skulls was found by Reverend William Fox (1813–1881), an eccentric clergyman who gossips would say prioritized fossil bones over his parishioners. Fox discovered a small, fairly complete skull of a herbivorous dinosaur on a beach on the Isle of Wight. Soon after, the famous anatomist Thomas Huxley (1825–1895) studied this specimen, and he recognized it as the skull of a small ornithopod, which he named *Hypsilophodon*. But the *Hypsilophodon* skull was unique, complete, and had a possessive owner in Reverend Fox. No one thought of sawing it in two or reducing it to dust in order to study its brain, which for the moment was of no interest to anyone.

In September 1869, still on the Isle of Wight and still on the beach, a friend of the good Reverend Fox came upon the back part of a mysterious skull on the foreshore. It had been polished by the beating of the waves and opportunely had broken to reveal the cranial cavity. The skull was soon examined by two of England's leading paleontologists—the discoverer himself, John Whitaker Hulke (1830–1895), and his friend Thomas Huxley. They were looking at the then-unknown *Iguanodon*, one of the first two dinosaurs to have been scientifically described almost fifty years earlier. Hulke took his time to describe the major areas of this paleo-brain in detail, noting various similarities with the brains of birds.[2] It was a pet project for Huxley, who was Charles Darwin's biggest supporter, to demonstrate the relationship between dinosaurs and birds. Although he convinced a few contemporary paleontologists, this remarkable concept, which was supported by a wealth of anatomical details and happened to be perfectly accurate, was sadly forgotten for another century. Hulke, a surgeon by trade with a specialty in ophthalmology, was certainly the right man to take charge of the first description of a dinosaur brain. In his younger days, his first claim to fame came in 1852, when he signed the death certificate for the Duke of Wellington, the man who had defeated Napoleon at Waterloo in 1815. Hulke then served as assistant surgeon for the British armies during the Crimean War, where he had some dangerous adventures, narrowly escaping Russian guns.[3] Even though he is barely known to anyone but paleontologists today (and no doubt to a few erudite ophthalmologists), Hulke was one of the key figures in dinosaur paleontology in the 1870s and the pioneer of paleoneurology. If the good doctor is well forgotten today, it may be because two of his contemporaries from across the Atlantic have taken pride of place in the history books of paleontology, not just for their bad manners but also because of the importance of their discoveries: Othniel Charles Marsh and Edward Drinker Cope, the famous duo behind the infamous Bone Wars.*

---

* Marsh and Cope were pioneers of vertebrate paleontology in the United States. From the 1870s onward, they were constantly and ruthlessly competing, both in the field, where their teams spied and played nasty tricks on each other, and in the columns of scientific journals.

And One, and Two, and Three Brains

A few years later, it was this very American paleontologist, O. C. Marsh, another acquaintance of Huxley's, who studied the endocranial casts of several dinosaur species brought back from the Wild West by his discovery teams. In 1880, he became the first paleontologist to examine the brain of *Stegosaurus*, a Jurassic dinosaur he had named just three years earlier. Marsh briefly described the main features of *Stegosaurus*'s endocranial cast. He thought he saw large optic lobes and reduced hemispheres. By comparison, when Marsh dissected a young alligator, he measured its brain and found it was about ten times smaller than the *Stegosaurus*'s and had a body mass a thousand times less (the alligator was really tiny). He concluded that a *Stegosaurus* of the same mass as this little alligator would have had a brain a hundred times smaller. This was a major calculation error, as we'll see later, since brain size doesn't increase in proportion to an animal's mass. In fact, the relationship between the size of the brain and that of its owner is allometric, meaning that a small animal has a proportionally larger brain than a large animal. Comparing stegosaur and alligator brains directly with cross-multiplication, as Marsh did, therefore has no biological relevance.

It is true, however, that *Stegosaurus*'s brain is not particularly impressive—it's one of the smallest dinosaur brains. Had Marsh left it at that, *Stegosaurus*'s reputation would have been set as king of the Mesozoic morons. Alas, alas, alas, he made another error the following year after discovering that *Stegosaurus*'s spinal cord widened considerably at the sacral vertebrae. The neural canal (through which the spinal cord passes) measured 12 centimeters (nearly 4.75 inches) in diameter in places, or three times the diameter of the brain. When Marsh compared the endocranial cast with the spinal-cord cast from the pelvic area, he noticed a vague resemblance in shape between the two. He referred to this cavity in the sacral vertebrae as "a posterior brain-case,"[4] which can be interpreted to mean either "a posterior cavity for a brain," which is completely idiotic, or "a cavity for a posterior brain," which is no better. Marsh suggested the existence of a posterior nerve center "at least ten times" larger than the brain. This accurate observation (the neural canal in the sacral vertebrae of *Stegosaurus* is indeed enlarged, and the volume of this cavity is much greater than that of the brain) quickly gave rise to the popular misconception that *Stegosaurus* had a "second brain" that compensated for the first brain's obvious deficiencies. The urban legend of the second brain was born, and it's still circulating today at the beginning of the third decade of the 21st century. As we shall see, the legend of *Stegosaurus*'s "two brains" was repeated ad nauseam with a few colorful variations over the next century and a half.

A quarter of a century after the publication of Hulke's first article, Marsh featured at least seven dinosaur brains in his work "The Dinosaurs of North America," published in 1896.[5] Marsh, who couldn't care less about museum visitors, had no qualms about cutting through skulls with a saw to see what was inside. In fact, this method enabled him to illustrate more dinosaur brains than any of his successors over the next few decades. Marsh was also an adamant evolutionist and, like Huxley, considered birds to be close relatives of dinosaurs. He illustrated the brains of different species, but he didn't dwell on the biological implications of the obvious differences he found. He was in the last years of his life and perhaps didn't have the time to focus on this aspect of his discoveries, preferring instead to be pleased with the accuracy of the "law" of brain evolution he had formulated in the 1870s. (Like most biological "laws" formulated in the 19th century, Marsh's law has since been relegated to the status of quaint historical harebrained ideas.) As a skilled anatomist, however, he noted the small brain size of the large sauropods *Camarasaurus* and *Diplodocus*, and the large olfactory lobes of the hadrosaur *Claosaurus*.

Fifteen years after dropping the "second brain" bombshell, Marsh revisited the topic of the enlarged spinal cord in the pelvic region noted in *Stegosaurus* and in sauropods, though he drew no definitive conclusions. At this point he claimed the volume of the pelvic cavity was "at least twenty times" that of the brain. Marsh again referred to a "posterior nervous center" controlling the movements of the posterior part of the body, noting this was "a suggestive subject, which need not here be discussed." As he died three years later, he didn't pursue the topic at all, and perhaps that's for the best. Although, others took up the task, to *Stegosaurus*'s great misfortune: "All dinosaurians are characterized by a very small brain, smaller than in any other known animal [body] form. Marsh, comparing the brain of a species of Dinosauria with the brain of a present-day alligator, found that the dinosaurian brain was a hundred times smaller than the alligator brain proportionately speaking (meaning by mathematically reducing the two animals to the same volume)."[6] These were the words of French paleontologist Marcellin Boule (1862–1941) in 1891, who had just returned from a trip to the United States, delighted and enthralled by Professor Marsh. Note the slightly outrageous generalization of "all" dinosaurians being stupid! A few years later, Boule drove the point home, writing,

> Dinosaurians represent the reign of brute force; these animals, with their imposing mass, must have had a slow, heavy gait. They were unintelligent beings, judging by their brains, which were very small compared to the size of their heads. Marsh has calculated that, proportionately speaking, the brain of a crocodilian living today, which cannot be considered to be a very intelligent animal, is a hundred times larger than the brain of a *Brontosaurus*! One can see that the spinal cord [of *Stegosaurus*] was much larger than the brain.... *Stegosaurus* had more intellect in its back than in its head![7]

WHAT DID DINOSAURS THINK ABOUT?

It's a mystery what Boule had against the poor dinosaurs, but they were the target of his hateful criticism for years. At an official ceremony on June 15, 1908, where American billionaire Andrew Carnegie donated a reproduction of a *Diplodocus* skeleton to France, Boule spoke these definitive words to French president Armand Fallières, as reported by the French daily newspaper *Le Matin*: "That animal was an imbecile. His brain was no bigger than a chicken egg." Redemption for *Stegosaurus* was not to be achieved by Boule, nor by his German colleagues for that matter. In 1915, the German paleontologist Edwin Hennig (1882–1977) described another stegosaur and gave it the gentle name of *Kentrosaurus*, meaning "prickle lizard." This cousin of *Stegosaurus* from the Jurassic Period was discovered by German expeditions in Tanzania just before the First World War when the country was a German colony. Like Marsh, Hennig made side-by-side illustrations of the endocranial cast of his dinosaur, which was in rather poor condition, and the much larger cast of the pelvic cavity without even attempting to elaborate on this touchy subject. By 1915, the depiction of *Stegosaurus*'s brain and the spinal-cord cast had already become a meme (a viral image), which incidentally shows Professor Marsh's influence long after his death. While Hennig remained cautious—silent even—on the subject, unfortunately other German anatomists greedily seized the window of opportunity. In 1914, one of them claimed the pelvic cavity contained a second brain in charge of digestion and reproduction! The American paleontologist Richard Swann Lull debunked this surprising hypothesis in 1917,[8] pointing out that since the vagus nerve controls digestion in crocodiles and the orifice of the vagus nerve is present in the cranium of dinosaurs, there can be no doubt that it fulfilled the same function in dinosaurs as in all other amniotes. Lull went on to point out two significant enlarged areas of *Stegosaurus*'s spinal cord: the one in the pelvis that Marsh and Hennig had noticed, and a second one in the shoulder region, both of which were related to the nerves departing toward the hind limbs and forelimbs, respectively.

In the meantime, a few other dinosaur brains had been described. In 1912, another American paleontologist, Henry Fairfield Osborn, illustrated the brain of his favorite dinosaur, *Tyrannosaurus rex*. To accomplish this, with a saw he cut open one of the hind skulls discovered in Montana by his legendary fossil hunter Barnum Brown.[9]* Cautious, Osborn merely described the cast of the cranial cavity, stressing that the brain itself would have occupied no more than half of this volume, around 250 cubic centimeters (about 15.25 cubic inches). "The excessively small size of the brain, probably weighing less than a pound, which is less than 1/4000 of the estimated body weight, indicates that an animal's mechanical evolution is quite independent of the evolution of their intelligence; in fact intelligence compensates for the absence of mechanical perfection." This was Osborn's way of talking, as he too liked to search for important laws of evolution, which, as successful as Marsh's "laws," also justly fell into oblivion.

As for Barnum Brown, in 1914 he described the brain of *Anchiceratops*, a cousin of *Triceratops*. After completing incredible fossil preparation work, which involved removing the bone to preserve the infilling of the cavities (brain, cranial nerves, and inner ear), the result appeared to have come out of a CT scanner, an invention that wouldn't exist for another three-quarters of a century.[10] As Brown remarked, the semicircular canals of a dinosaur's inner ear had just been discovered for the first time.

Most of this research remained overlooked, however, because it was overshadowed in both popular and scientific publications by

---

* Barnum Brown remains the greatest dinosaur hunter in the history of paleontology. In his fifty-year career with the American Museum of Natural History, he brought back most of the skeletons still on display today in this great museum.

Stegosaurus's "two brains." Symbolic of dinosaurian stupidity, it was believed that this supposed lack of intelligence must have been the cause of their extinction. Animals with such small brains were bound to become extinct to make way for far more cerebrally equipped beings: mammals. The culmination of the two-brain myth was probably reached and even surpassed in 1956 by Richard Carrington, an author of popular books. He asserted, "*Stegosaurus* had at least three brains,"[11] one in its head, one in its shoulders, and one in its hindquarters.

German American researcher Tilly Edinger (1897–1967) is often considered to be the true founder of paleoneurology. She was one of few women in the small, male-dominated world of vertebrate paleontology in the mid-20th century. After a brilliant start to her career at the Senckenberg Museum in Frankfurt am Main, Edinger was thrown out in 1938, a victim of the Nazis' anti-Jewish laws. Most of the Jewish scientists had long since been driven out of their jobs or had already emigrated by 1938. This reprieve for Edinger was due to the museum being a private institution, her being a volunteer there, and her boss liking her. With a paleontological sense of humor, she considered herself "an ammonite in the Holocene" (ammonites disappeared 66 million years ago while the Holocene Epoch, the last epoch of the Quaternary Period, began 11,700 years ago). She refused to consider leaving Frankfurt, saying, "My mother's family has been here since 1560, I was born in this house. And I promise you they will never get me into a concentration camp. I always carry with me a fatal dose of veronal."[12] After Kristallnacht in November 1938, however, her situation became untenable. At the last moment, Edinger managed to escape to England in May 1939, then to the United States thanks to the intervention of the great American paleontologist and zoologist Alfred Sherwood Romer. She then began a second career at Harvard University's Museum of Comparative Zoology. Edinger spent her professional life studying the fossil brains of all kinds of creatures, which incidentally shows just how little interest there was in paleoneurology at the time, because the countless males dominating the profession would never have left an interesting subject to one of the very few female paleontologists. In fact, now that paleoneurology is fashionable, it's mainly practiced by men. But Edinger debunked the legend of *Stegosaurus*'s second brain,[13] explaining the enlarged neural canal to essentially be due to the development of nerves supplying the tail and limbs. More recently, it has been suggested that this enlarged region, also present in birds, was used to store glycogen, the function of which is not clear. It's highly likely that *Stegosaurus* also stored glucose in the pelvic area. What's more, gigantic sauropods also exhibit an enlarged neural canal in the same place, probably for the same reasons.

Measuring Intelligence

The natural temptation to define and quantify intelligence, whether human or animal, has consumed generations of psychologists, neurologists, and ethologists, with little consensus on the results. Brain size was long considered the best measure of human intelligence, an idea subtly illustrated in the 1969 film *The Brain*, directed by Gérard Oury, in which a genius burglar, nicknamed the Brain, played by David Niven, is so smart he has trouble keeping his head upright because of the weight of his brain! Many illustrious figures of the late 19th century requested their brains be weighed after their death in a kind of posthumous competition to see whose would be the heaviest. As it turned out, there was no evidence the weight of the illustrious departeds' brains reflected their intellectual performance. If I may sum it up this way, Einstein's brain is hardly any different in weight or size from that of any other random person. The scientific rigor that drives us should therefore banish popular expressions like "pea-brain" and "birdbrain" from our vocabulary.

The concept that brain size is directly related to intelligence posed another unpleasant problem for humankind when our brain mass was compared to that of other animals. While humans' average brain mass is around 1.3 to 1.5 kilograms (2.9 to 3.3 pounds), elephants' reaches 5 kilograms (11 pounds), blue whales' more than 7 kilograms (more than 15 pounds), and sperm whales' 9 kilograms (nearly 20 pounds)! Well, we couldn't accept that these animals might be intellectually better equipped than us, so to get out of this impasse and put the human species back in its rightful place (first place, of course), encephalization quotient was invented. Starting in the 1950s and 1960s, researchers began trying to quantify the relationship between brain size in living animals and fossil animals. The overall idea was to measure as many brains as possible and establish a relationship between the mass of the brain and the mass of its owner. When enough species had been measured in this way, the researchers were able to generate a scatterplot and use mathematical tools to draw a trend line that best fit the datasets. In paleontology, Harry J. Jerison, an academic from California, pioneered this approach in 1973.[14] Jerison, who had been in close contact with Tilly Edinger when he'd begun his quantification research, dedicated his book on the evolution of the brain to her. The book was published a few years after her accidental death. In his work, Jerison explained that as brain size increases with animal size, the encephalization quotient (EQ) can be used to determine relative brain development in different species. The EQ value is greater than 1 if the individual has a relatively larger brain than expected for an animal of its weight, and it is less than 1 otherwise. In other words,

EQ = actual brain mass / predicted brain mass.

Among mammals, the encephalization quotient of *Homo sapiens* is 7.4, dolphins 5.3, and rabbits 0.4.[15] So the rabbit's brain is two and a half times smaller than would be expected in a mammal of its size, while the human brain is almost seven and a half times larger than would be expected. It's true that no one claims the intellectual prowess of rabbits is superior to that of humans, apart from a few poets perhaps. (As French singer Chantal Goya sang, "This morning, a rabbit killed a hunter.")

Encephalization Quotient and Its Variants

As brain size increases with animal size, the encephalization quotient can be used to determine relative brain development in different species. The encephalization quotient for all species is calculated in the same way as for mammals (MEQ, for mammal encephalization quotient), but a different coefficient is used for reptiles (REQ, for reptilian encephalization quotient) and another for birds (BEQ, for bird encephalization quotient). Of course, many researchers have studied the REQ and BEQ of dinosaurs, but there are two additional challenges: estimating the brain weight based on the endocranial volume and weighing a living dinosaur—or at least estimating its weight, which is no easy task.

To determine endocranial volume, all you need are dinosaur skulls and an X-ray microtomography scanner, more commonly known as a CT scanner. The machine is kind of like a large microwave oven weighing a few tons that, rather than spitting out microwaves, bombards its target (like a dinosaur skull) with X-rays. It uses the same principle as medical scans, except the subject is blasted for an hour at maximum power. A patient's chances of survival would be slim at this power level, but this strength is essential to get through a fossilized skull. Just like in a microwave oven, the skull is slowly rotated on a small tray, which allows the scanner to create a few hundred cross sections of the skull's interior. These slices can then be assembled using a software program to create a 3D reconstruction.

Because the bone density and the density of the rock matrix that filled all the cavities are different, the details of the cranial cavity can be reconstructed and the cavity's volume precisely determined.

In general, it's estimated that a dinosaur's brain filled 50% of the cavity's volume (the meninges, arteries and veins, and interstitial tissue occupied the remainder). This estimate is corroborated by recent studies on the size of the crocodilian brain in relation to the size of its brain cavity.[16]

The question of dinosaur mass is another matter altogether. The overall idea is that by measuring the volume of a reconstruction of an animal, its mass can be deduced by presuming its density is the same as that of present-day animals (around 1). Of course, if you model an emaciated *Diplodocus*, you won't get the same result as with an overweight one. The margin of error in the final result is obviously considerable. For example, the same *Giraffatitan* skeleton analyzed by different researchers yielded a mass ranging between 15 and 78 metric tons. With such a wide range of results, it's fair to question the reliability of the whole exercise! A 2014 study[17] used the circumference of the femurs and humeri to deduce a dinosaur's mass. So, with a tape measure and a calculator, we can estimate that *Giraffatitan* weighed 56 metric tons, which seems consistent with estimating its weight by immersing a model of this animal in water.

Nonetheless, once all these calculations have been made, we obtain the dinosaurs' REQs, which inevitably reflect something about their intellectual capacities. Please note we are talking about REQs here, which are in no way comparable to the MEQs of mammals! So *Stegosaurus* (REQ between 0.37 and 0.47) rivals *Diplodocus* (REQ between 0.35 and 0.45) at the back of the pack. Therefore *Stegosaurus*'s brain is 0.37 to 0.47 times the size we'd expect from a reptile of its weight. Which isn't much. Among the poor performers are other sauropods (*Giraffatitan*: REQ between 0.52 and 0.90), ankylosaurs (*Euoplocephalus*: REQ of 0.89), and ceratopsians

(*Triceratops*: REQ between 0.67 and 0.83). The worst performer in this last group is *Pachyrhinosaurus* with its REQ of 0.50. Well ahead of them rank the ornithopods (*Iguanodon*: REQ of 2.57; *Anatosaurus*: REQ of 2.36), whose brain size is twice that expected in reptiles of the same size. Their encephalization coefficient is comparable to that of large theropods such as *Tyrannosaurus* (REQ of 2.07 to 2.57) and *Allosaurus* (REQ of 3.3 to 4.3). Small predators like *Troodon* (REQ of 7.06), *Dromiceiomimus* (REQ of 7.14 to 8.60), and *Velociraptor* (REQ of 12) top the list! Carnivores, unsurprisingly, have relatively larger brains than herbivores. For hundreds of millions of years, it has taken more coordination to track prey than to choose the best fern. This would raise concerns about the distant intellectual future of a human species that has become vegan, if we didn't also remember that choosing food in the fruit-and-vegetable section or the meat department of the supermarket doesn't require very different brain skills.

A recent doctoral dissertation[18] has improved the estimation of the actual volume of the brain compared to the endocranial cast. The results are quite similar, although the rehabilitation of *Stegosaurus* (REQ of 1.36) and *Triceratops* (2.14) is dramatic, and the pachycephalosaur *Stegoceras* with its REQ of 6.28 deserves a shout-out. Finally, by far the worst performer of the group is *Diplodocus*, with a quotient of 0.36. The REQ of *Diplodocus* is therefore comparable to the MEQ of the rabbit, which probably means nothing: comparing an MEQ and an REQ doesn't tell us anything, and if we calculate the MEQ of a dinosaur, the cleverest of them all pales in comparison. To put it this way, the intellectual *Troodon* with a triumphant REQ of 7 only has an MEQ of 0.23. But getting back to our stegosaur, let's note that with its REQ of 1.36, it far exceeds that of the alligator, which stands at 1.05. Marsh's assertion that the alligator's brain is a hundred times larger than the stegosaur's was completely unfounded. While the alligator's brain size is close to average for a reptile of its size, the stegosaur was relatively better endowed.

The encephalization quotient therefore gives a first approximation of a dinosaur's cerebral capacity, with some dinosaurs appearing to be better performing than present-day crocodilians and approaching—or even surpassing—the performance of many birds. In this respect, calculating the REQ of present-day birds has shown that all have a value above 4, with some parrots having REQ values over 30. Although, there is the notable exception of the red-throated loon (*Gavia stellata*), whose REQ value is 1.20. At the risk of being harassed by an ornithological society, I suspect that the red-throated loon, which I'm not very familiar with by the way, is a bird with limited intellectual abilities. But the same cannot be said of all winged creatures: parrots and corvids alike are remarkably intelligent and have relatively large brains. And, as in so many other areas, (brain) size isn't everything. Biologists have recently demonstrated that the number of neurons in the forebrain (the prosencephalon) is the best approximation of intelligence, and the neuronal density of some birds is much higher than that of mammals.[19] Unfortunately, the neuronal density of dinosaurs is still inaccessible to us.

It's perhaps not completely implausible, however, that dinosaur neuronal density could be a topic we're able to discuss at some point in the future. In theory, a brain can't be preserved, but paleontology is full of exceptions, which is part of its charm. British researchers have described a natural endocranial cast of an *Iguanodon* found on a beach in Sussex. Yes, English beaches have been endless sources of dinosaur skulls since 1869, perhaps because if you can't put your toe in the icy water, there's nothing to do but explore the allure of each pebble. A natural endocranial cast occurs after an animal's death when sediment fills the cranial cavity; the skull bones almost entirely erode away, leaving only what appears at first glance to be rock. But first impressions can be deceiving, so these British researchers upgraded to scanning electron microscopy and X-ray microtomography.[20] They believed they could distinguish the meninges and their blood vessels enveloping the gray matter of the cortex. These soft tissues would therefore have been partially fossilized in this specimen, paving the way for more detailed investigations in the future.

When it comes to intellectual performance, it's the small carnivorous dinosaurs that stand out. Theirs were the Rolls-Royces of dinosaur brains, with encephalization quotients (REQs) as high as 7 to 12, and large cerebral hemispheres. By comparison, their brains were considerably larger than crocodilians' but much smaller than parrots' (*Troodon*'s REQ is 0.63 compared to 2.36 for *Psittacus erithacus*, the gray parrot, a parrot particularly gifted at oral expression). Canadian paleontologist Dale Russell envisaged the post-Mesozoic evolution of *Stenonychosaurus* in a world where dinosaurs had never gone extinct.[21] *Stenonychosaurus* was a small troodontid, a close cousin of *Velociraptor*. Perhaps taking inspiration from a certain primate species (humans), Russell commissioned the construction of a life-sized sculpture of a hypothetical distant descendant of *Stenonychosaurus* with a shortened tail and enlarged brain. In a word, he created a "dinosauroid," a thinking reptile with a big brain. The result of this amusing fictional evolutionary experiment was a bipedal humanoid dinosaur; dressed in a hoodie, you'd pass it (almost) without noticing it on the subway. In the wake of the "dinosauroid" and the various *Jurassic Park* raptors that become increasingly intelligent with each new film, the evolution of thinking raptors has become a science-fiction classic in recent years. In *Evolution*, novelist Stephen Baxter describes Jurassic theropods of the genus *Ornitholestes* as being equipped with whips and spears to attack *Diplodocus* and, incidentally, eliminate them. This has led even serious paleontologists to speculate on the possibility that thinking dinosaurs might have fashioned wooden tools that have since fossilized and could be identified.[22] Postulating that many birds, and even crocodilians, use tools leads us to ask why not consider dinosaurs? Just to clarify, while birds haven't created a wrench (yet), corvids can use twigs or iron rods to catch interesting objects.[23] To summarize D. J. Varricchio et al.'s enticing article with its slightly disappointing conclusions: no dinosaur tools have been found, and even if they were, it would be a challenge to identify them. The novelist Jean-Luc Marcastel takes up the same concept in his work of fiction *Tellucidar*, in which philosophical raptors are the intellectuals of a contemporary subterranean world.[24]

The other nerds of Dinosauria are the ornithopods, with respectable REQs and sizable cerebral hemispheres. Like Russell and his "dinosauroid," novelist Dominique Delpiroux has chosen the heroes of his novel *Les doigts du Diable*[25] (The devil's fingers) well, in which thinking hadrosaurs are among us. Hadrosaurs' cerebral hemispheres represent 45% of the endocranial volume, which is more than in large theropods, such as *Carcharodontosaurus* (24%) or *Tyrannosaurus* (33%), and approximately the same as in some small theropods, such as *Conchoraptor* (43%). The relatively large size of the hemispheres is perfectly in line with the complexity of the behaviors attributed to hadrosaurs, particularly in terms of vocal communication, which we will discuss later.

I don't know whether all these writers felt a sense of loss over the evolution of dinosaurs being so brutally interrupted by the impact of a meteorite and felt the need to correct this injustice. But in any case, many people believe natural evolution should, ultimately, produce a creature with large cerebral hemispheres and an impressive encephalization quotient. This is a horrifically anthropocentric and, of course, extreme view of evolution! Evolution doesn't have a goal; it's simply a set of phenomena that enable living organisms to adapt to changes in their environment. And the evolution of the human brain is no more or less likely than that of the eye of the trilobite.

Faced with the avalanche of thinking dinosaurs in literature, we must nevertheless raise a legitimate question. Despite what we've just said, if an intelligent species of dinosaur had existed, how would we know? Fortunately, exobiologists have been looking into the matter![26] Exobiologists (or astrobiologists) are specialists in extraterrestrial life, past (exopaleontologists) and present. As their object of study has yet to be discovered, they are testing their working methods on the only planet available to them at the moment: Earth. In the *International Journal of Astrobiology* (admittedly a rather frustrating journal), two American scientific scholars asked what geological traces an industrial civilization would leave millions of years after its disappearance. To answer this question, the authors examined our own civilization and its geological markers. What traces will we leave behind that will still be observable or measurable in a few tens of millions of years? Not buildings or manufactured objects that will have been long destroyed by erosion, subduction, and so forth. But certainly a spike in carbon dioxide (CO_2) (or at least its geological equivalent, an increase in the isotope carbon-12 compared with isotope carbon-13 in limestone), perhaps layers of plastic residues, and undoubtedly a spike in plutonium-244. Armed with this inventory, now we just need to explore Earth's geological layers in search of these three signatures to determine whether another civilization preceded us on our planet.

Is there any evidence of plastics in Earth's strata? No. A spike in plutonium-244? No again. A spike in CO_2? Oh yes, for example during the Paleocene-Eocene Thermal Maximum. In the span of perhaps five thousand years, a massive quantity of carbon dioxide was injected into the atmosphere, fifty-six million years ago! Could this be the signal we're hoping for? Could an intelligent species have evolved in the Paleocene Epoch, causing global warming just like the first *Homo sapiens*, before simply becoming extinct? Unfortunately, the abrupt hyperthermal events of the past, which are due to a sudden influx of CO_2 or CH_4 (methane) into the atmosphere, are generally contemporaneous with episodes of volcanic or tectonic overactivity. This suggests it was simply the intrusion of magmas into organic-rich shales (or petroleum-bearing evaporites) that led to the atmospheric outgassing of large quantities of greenhouse gases (CO_2 or CH_4). Warming during the Paleocene Epoch and other earlier periods is "thus not sufficient evidence for prior industrial civilizations."[27]

It follows from this interesting study that no dinosaur species (nor marsupial, nor squid) has ever established an industrial civilization. Even if we thought that conclusion might be true, we still prefer to have scientific confirmation.

While the examples of intelligent raptors and hadrosaurs are certainly inspired, Bertha, the large thinking brachiosaur from "Our Lady of the Sauropods," a short story by American science-fiction writer Robert Silverberg,[28] is far less convincing. In sauropods, it's not brain size that impresses; they have by far the lowest encephalization quotients and strangely organized brains. Their hemispheres are tiny, but the pituitary fossa beneath the brain is relatively large. The fossa contains many things, such as the cerebral carotid arteries, the oculomotor nerves, orbital veins, and the pituitary gland itself. The pituitary gland (or hypophysis) produces hormones, notably growth hormone. And if the largest land animals that ever existed had a large pituitary fossa, it's probably because their hypophysis was too.[29] No doubt their hypophysis had to produce a whole lot of hormone to achieve unrivaled growth.

An intriguing trend emerges in the evolution of the group, one that's bound to shock the big-brained synapsids reading this book: the oldest sauropodomorphs, creatures from the Late Triassic Period like *Buriolestes*, have a significantly higher encephalization quotient than their distant descendants from the Jurassic and Cretaceous Periods. *Buriolestes*'s REQ is 0.65 (two-thirds of what is expected for a reptile of its size), compared with 0.30 to 0.35 for large Jurassic sauropods. In other words, the relative size of its brain, not much to boast about from the start, decreased over the course of sauropodomorph evolution. The fact that they switched from eating meat to eating plants is certainly part of the reason why.

The Size of a Walnut

Ever since Marsh's distressing ramblings on the alleged stupidity of dinosaurs, a unit of measurement has gradually taken hold in popular science books, and that is the walnut. This may seem odd, given that we use units such as liters or gallons and cubic meters or cubic feet (in this case, milliliters or fluid ounces and cubic centimeters or cubic inches) to measure volume. But the walnut certainly has a more pathetic quality, and it's much easier to visualize a walnut than 26.2 cubic centimeters (1.6 cubic inches), the volume of an average walnut. In short, since the beginning of the 20th century, it has been written countless times that the stegosaur brain had the same volume as a walnut, and all the readers have sneered, troubled by the thought of an animal weighing several tons being guided by a walnut-sized brain. In 1996, Canadian paleontologist Grant Hurlburt estimated the volume of *Stegosaurus*'s cranial cavity to be 45 milliliters (1.5 fluid ounces), or almost two walnuts, but given the presence of meninges and so forth in the cavity, he concluded that only 50% was occupied by the brain, or 22.5 milliliters (0.8 fluid ounces).[30] So indeed, pretty much a walnut. Thanks to the arrival of CT scans, more precise calculations of the stegosaur brain volume have been made more recently. The volume of the cranial cavity of an average *Stegosaurus* would in fact be 93 cubic centimeters (about 5.7 cubic inches) and the volume of its brain around 47 milliliters (nearly 1.6 fluid ounces). That makes almost two walnuts. Or a beautiful apricot, if you prefer.

3

In Search of Lost Senses

The recent investigations into the brains of dinosaurs allow us, for the first time, to scientifically ask questions that no paleontologist would have been willing to answer until now: what did dinosaurs see, smell, and hear? While we still can't know their thoughts, it's now possible to at least try to analyze how they conceived the world in terms of their cerebral capacities. The many dinosaur skulls that have spent time in CT scanners in recent years have provided various clues as to dinosaurs' sensory acuity. According to 21st-century biologists, the five classical senses defined by Aristotle (sight, smell, hearing, taste, and touch) are slightly outdated when it comes to the animal world. In reality, the senses fall into broad categories, depending on how they function. Chemoreception concerns everything to do with the molecules perceived by the brain, whether they float in the air (smell) or are ingested (taste). Mechanoreception includes touch but also hearing, and involves the perception of sound waves transmitted through the air or ground, and sometimes echolocation. Thermoreception is the skin's sensation and perception of temperature. The senses related to the reception of electromagnetic energy are sight, electroreception (the ability to perceive electric fields, especially in certain species of fish), and magnetoreception (the ability to detect variations in the earth's magnetic field, which is common in birds). And lastly, nociception is the perception of a painful or injurious stimulus.

"All animals, from doodlebugs to elephants to humans, 'inhabit' different sensory worlds," wrote biologist Benoît Grison. "And in terms of adaptation, it cannot be said that one of these 'perceptions of the world' is more 'accurate' or less 'relevant' than the others."[1] Even if studies are still mostly incomplete (only a few dozen dinosaur brains were studied in 2022), it's already clear that each of the tens of thousands of dinosaur species that existed had its own conception and experience of the world: a field of scents for one, a medley of colors for another, a symphony of sounds for yet another, and, needless to say, endless combinations of all those sensations.

The Scent of Hadrosaurs in the Air

Some *Homo sapiens* perceive the subtle scent of banana in Beaujolais nouveau wine and aromas of hazelnut in Bordeaux. But neither the best enologist nor the best perfumer has ever been asked to pick up the scent of a wild boar or to roam an oak grove on all fours to find the best truffles. For that there are dogs, whose sense of smell is incomparably superior to our own. A Labrador retriever has 150 million olfactory receptors while our species has only 5 million—game over. But what about the olfactory acuity of the characters in *Jurassic Park*? The film's paleontologist, Alan Grant, advises staying put to escape becoming *T. rex*'s prey, because supposedly this dinosaur can only detect moving objects. Was this the right strategy? Did *Tyrannosaurus* really lack olfactory reception? In the real world, would this advice have helped Tim and Lex dodge an attack?

To answer these empirical questions, in 2008 Canadian and Japanese researchers turned their attention to the anterior end of carnivorous dinosaurs' braincase, where the olfactory bulbs that process odor reception are located.[2] To compare different species of theropod dinosaurs, they measured the olfactory ratio, or the ratio of the greatest diameter of the olfactory bulb to the greatest diameter of the cerebral hemisphere. Analogous with present-day animals, they assumed that the higher this ratio, the more important the sense of smell was in a dinosaur's life. The researchers' results put *Tyrannosaurus* well ahead of the pack, with significantly larger olfactory bulbs relative to predicted values; allosaurs had olfactory ratios close to predicted values; and ornithomimosaurs were at the other end of the scale, with significantly smaller olfactory bulbs for theropods of their body size. Clearly the sense of smell must have been an essential part of tyrannosaurs' daily life.

By comparing the results obtained for theropods with what is known about present-day birds and mammals, some additional insights emerge. The development of the olfactory lobes in the brain is often linked to locating prey in low-light conditions, such as at dusk or at night, but it can also be correlated with locating prey that is difficult to see in all lighting conditions (for example, animals that use various types of camouflage). In carnivorous mammals, home-range size is often also correlated with olfactory-bulb size: the larger the animal's home range, the larger its olfactory bulbs and greater its olfactory acuity are. These observations suggest a number of hypotheses for tyrannosaurs, all of which are dismaying for their fellow creatures: they could detect prey whatever the light conditions, camouflage wasn't enough to protect their potential prey, and it's possible they evolved over a vast home range.

In 2019, researchers at University College Dublin took this work one step further.[3] They established a correlation between the olfactory-bulb ratio and the number of olfactory receptor genes in different present-day species of birds, mammals, and crocodilians. Olfactory receptor genes activate the production of proteins that act on olfactory receptor neurons that then signal the olfactory bulb, enabling the brain to transform airborne odorant molecules into odor perception. Put simply, the larger the olfactory bulbs are in relation to the rest of the forebrain, the more olfactory receptor genes there are and the better the sense of smell. Among present-day animals, elephants are the world champions with over two thousand olfactory receptor genes (over four thousand counting pseudogenes, which are nonfunctional genes), while *Homo sapiens* possess only a paltry four hundred functional olfactory receptor genes.

The researchers extrapolated results for various dinosaurs from the data they collected for living animals, inferring the number of olfactory receptor genes from their olfactory-bulb ratios. As dinosaur genes have not been conserved over millions of years (the oldest known pieces of DNA are only around one million years old), this is the only approach currently possible. What's more, we can even get an idea of the nature of these olfactory receptor genes, since those common to present-day crocodilians and birds inevitably existed in dinosaurs! This important study found that *T. rex* had the largest repertoire of olfactory receptor genes, estimated at 645 genes, considerably more than most modern birds have and close to what a dog has (800 genes).

To sum it up, *Jurassic Park*'s characters didn't stand a chance against a "real" *Tyrannosaurus*. Staying still all night would have been of little use, as *T. rex*'s exceptional sense of smell would infallibly guide it to its prey. Alan Grant would have led our two protagonists to certain death!

What about other dinosaur species? Many of them had large sinuses. Sinuses are cavities in the bones of the skull that connect to the nasal cavity. Humans have sinuses, but in dinosaurs, sinuses are hypertrophic, or abnormally enlarged, and take up a significant part of the skull bones, which must have resulted in sinusitis like we can't imagine. While these sinuses more likely had a role in thermoregulation than in the sense of smell, all dinosaurs also had developed olfactory lobes, but to varying degrees. Ornithomimosaurs and oviraptorosaurs had distinctly smaller olfactory lobes than other theropod dinosaurs, which has been interpreted as a further indication that they were likely omnivores, despite the fact that present-day animals show that herbivores (such as elephants, cows, and horses) have the largest olfactory-receptor-gene repertoires.

Up to now, detailed studies of the olfactory bulbs in the forebrain have focused on carnivorous dinosaurs. Data for other dinosaur groups is more fragmented. Most sauropods had large olfactory bulbs, so they must have been sensitive to odors, which may have been related to their sociality (sauropods lived in herds) and need to discriminate between the odors of plants to discern which were edible. On the other end of the spectrum we have the ceratopsids. *Triceratops* and *Pachyrhinosaurus* had relatively small olfactory bulbs and must have had reduced acuity in their sense of smell. They may have relied more on their sense of hearing to perceive the world.[4] Since they likely couldn't smell themselves, they might have used their own stinky odor to ward off predators sensitive to bad smells. In contrast, the small marginocephalian dinosaur *Psittacosaurus* from the Lower Cretaceous of Asia had large olfactory bulbs. This is also the case for representatives of the other group of marginocephalians, the pachycephalosaurs. While pachycephalosaurs' skulls could reach up to 20 centimeters (almost 8 inches) in thickness, giving them a false air of intelligence, the purpose wasn't to house a big brain. Under this thick, compact bone was a small cranial cavity, similar in size to that of other dinosaurs. In any case, their olfactory bulbs were relatively large, a sign that *Pachycephalosaurus* and its cousins had a good sense of smell.

The rare ankylosaur skulls that have been CT scanned, such as *Euoplocephalus*'s skull, also show relatively large olfactory bulbs[5] along with hypertrophic nasal cavities. Odors were therefore an important part of their existence. In *Ankylosaurus*, the giant of the family measuring 8 to 10 meters (26.25 to 32.8 feet) long, the nostrils are slightly offset to the sides of the skull rather than forward facing.[6] This arrangement could have given *Ankylosaurus* a convenient stereolfaction for orienting itself according to the dominant scents. The same hypothesis has been put forward for present-day moose (*Alces alces*), which also has widely separated nostrils.

In short, most dinosaurs seem to have had a well-developed sense of smell for sniffing Mesozoic odors, about which we know very little. Bakeries didn't exist yet, and fragrant flowers didn't appear until the Cretaceous Period, but the pungent vomit-like odor of ginkgo ovules existed as early as the beginning of the Triassic Period, as did the bouquet of conifer resins. As for the odors of the dinosaurs themselves, we don't know what they were; but we do know that crocodilians are the proud owners of paracloacal glands (located inside their cloaca, the rear orifice) and that birds have uropygial glands (at the base of the tail), which suggests that dinosaurs must also have had musk glands in the cloaca. These glands produce a wide variety of odors, not necessarily putrid ones. For example, caimans secrete citronellol, an organic compound with a roselike odor that can also be obtained by distilling lemongrass. I realize, however, that I've been a little hasty in passing over what may only be a detail for you: how did we know the caiman's paracloacal glands smelled of roses? For a brief moment, the image passed through my mind of a scientist crawling through the mud toward a caiman in the heart of Mato Grosso, Brazil, and delicately inserting a pocket spectrometer into the animal's cloaca. But as zoologists specialized in crocodilians have a life expectancy quite comparable to that of the rest of the population, I felt it necessary to reread in detail the article from which I had taken the information. So I went back to "The Chemistry of Crocodilian Skin Glands" by

Paul J. Weldon and James W. Wheeler,[7] assuming I'd discover a more sinister reality: execution, dissection, chromatography, and all the rest. But no, the sample is taken from living animals—in captivity, not in the wild—by manually palpating and compressing the glands to collect the secretions. Getting back to the subject at hand, did tyrannosaurs smell like lavender or give off a stench? For the moment, we don't have an answer to this legitimate question, but that may not always be the case thanks to the recent emergence of molecular paleontology, which searches for organic molecules in increasingly ancient fossils. In 2020, the presence of melanin was demonstrated in the cloaca of a 125-million-year-old *Psittacosaurus*.[8] Only the outer part of this small dinosaur's cloaca was preserved and, incidentally, was the first specimen of its kind ever discovered. The discovery of a dinosaur's cloacal glands has yet to be made, but it could one day shed light on their body odors.

In addition to the supposed odors of their supposed cloacal glands, it would be unfair not to discuss in this olfactory section the possibility that dinosaurs farted. A fart doesn't fossilize, fortunately, but the production of methane during digestion leads many animals to expel this gas from their anus, which is both noisy and smelly (or through their mouth, by burping). Birds generally do not fart, as their digestion is extremely rapid. On the other hand, some snakes flatulate, crocodiles sometimes do too, and dinosaurs were probably the worst offenders, contributing to greenhouse gas emissions like cattle do today. By supposing that sauropods had a long digestion process involving the production of gas, researchers calculated they would have produced 500 million metric tons of methane every year (ten times more than cattle do), contributing to the warm Mesozoic climate.[9] To arrive at this result, they estimated the sauropod population and biomass density to be approximately ten 20-metric-ton individuals per square kilometer (0.4 square miles) spread over half the total land area, each emitting 2,675 liters (almost 707 gallons), equivalent to 1.9 kilograms (more than 4 pounds), of methane per day. To be honest, we don't know anything about sauropods' gut microbiota, the gases produced by their digestion, or their population density. At best, this is a rough estimate.

Rereading these lines, I think it's fair to point out that odors in the Mesozoic Era must have included more than just vomit and methane fumes; the introduction of flowering plants during the Cretaceous Period must have brought new and more pleasant fragrances, the sea air was already iodized, and many other pleasant scents floated in the atmosphere. As for dinosaurs' cloaca, I can't help but think perhaps it secreted some bewitching perfumes.

See *T. Rex* and Die

When we scan a dinosaur skull, we obtain a 3D image of the animal's cranial cavity, which contained not only the brain but also veins and meninges. These layers make it difficult to recreate in 3D the structures located behind the cerebral hemispheres. Unlike the olfactory lobes, the optic lobes are also complicated to reconstruct. Another element, however, appears more easily: the flocculus. The flocculus is a small cerebral mass located behind the cerebral hemispheres that serves to stabilize the head and maintain a steady image on the retina when the head is moving. Modern large apterous (wingless) birds are known to have large flocculi, suggesting this is also a bipedal adaptation to help stabilize the inherently unstable nature of bipedalism. There's a reason why a biped whose facultics have been impaired by the ingestion of psychotropic drugs ends up on all fours. The flocculus plays a part in the vestibulo-ocular reflex, which means that when you turn your head to follow a moving object, your eye muscles automatically act to stabilize the object's image on your retina. But that's not all. The flocculus performs another important reflex—the vestibulocollic reflex—which stabilizes the head in relation to the center of gravity when the gaze is fixed and the body is in motion.

In the absence of well-defined optic lobes, however, relative visual acuity of different dinosaurs can't be estimated in the same way as olfaction because it's impossible to measure the portion of the brain devoted to sight. Even so, we do have some skeletal information, including the size and position of the orbits, the development of the flocculi, and the semicircular canals of the inner ear. No, you're not dreaming; we can measure the size of the semicircular canals in the inner ear of dinosaurs, and it's amazing. Remember there are three canals, each detecting respective head movements: vertical, when you move your head up and down (anterior semicircular canal); lateral, when you turn your head side to side (lateral semicircular canal); and in the coronal plane, when you tilt your head toward your shoulders (posterior semicircular canal). The vestibulo-ocular reflex is closely linked to the development of the semicircular canals. For example, in visual predators (but also in agile arboreal species), these canals are relatively large, while animals for which vision is less essential, such as cetaceans, have smaller canals.

It turns out that, in tyrannosaurs, the lateral canal is particularly elongated, suggesting the importance of lateral head and eye movements,[10] and this shape probably wasn't to help them watch a game of tennis! The considerable size of semicircular canals in *Tyrannosaurus* and many other theropods is clearly linked to their predatory habits.

In contrast, the semicircular canals of sauropods are often atrophied, which is not unexpected in herbivores[11] as rapid head and eye movements weren't their thing. And acrobatics weren't their thing either, which should come as no surprise. While no one ever hypothesized that sauropods were arboreal creatures leaping from branch to branch, the discovery of their inner ear confirms this hypothesis would have been wrong. Even so, they weren't nearsighted. Their orbits are large, so they must have had big eyes, as evidenced by the scleral rings found on several specimens. The scleral rings found in turtles, lizards, and birds are small articulated ossicles that form a circle around the cornea. The size of the circle they form approximates the diameter of the eye. Sauropods, therefore, had good eyesight, but they simply couldn't turn their heads too quickly or else everything would blur.

In addition to their size, another distinctive feature of dinosaurs' orbits is their arrangement. When they're facing forward, like *Velociraptor*'s and *Tyrannosaurus*'s, a wide binocular field of view is possible (meaning both your eyes see the point of view in front of you simultaneously), which is useful for a predator. This arrangement in particular enables stereoscopic vision with better depth perception. But when the orbits are located on opposite sides of the head, like in cattle, rabbits, and *Diplodocus*, the binocular field of view is almost nonexistent.

MIDNIGHT DEMONS

The fossilization of scleral rings has led to another hypothesis put forward by researchers at the University of California.[12] According to them, the diameter of the rings in question and the size of the orbits within which the rings are located are correlated with the nocturnal or diurnal activity of their owners. Diurnal animals should have smaller scleral rings that don't completely fill the orbital cavity, whereas nocturnal animals should have large scleral rings that completely fill the orbit with a considerable inner diameter. As all photographers will understand, this helps to let in more light at night to get an image. In any case, the researchers' results suggest that small predators such as *Velociraptor*, *Juravenator*, and

Microraptor were scotopic (nocturnal), while almost all herbivores were active during both day and night. So these predators hunted by moonlight.

In addition to large scleral rings compatible with nocturnal activities, *Velociraptor* also possessed relatively large semicircular canals and, above all, enormous flocculi, the largest known in a dinosaur.[13] It also had a large braincase, large orbits, and stereoscopic vision; or in other words, it had all the attributes of a lively, fast-moving predator capable of pursuing prey by day and night and changing direction quickly—truly vicious!

Shuvuuia[14] also had large scleral rings that occupied the entirety of its large orbits and was certainly a nocturnal animal. This small predator from the Alvarezsauridae family possessed a quite astonishing characteristic known only in a few members of this family of small theropods, and that was a single functional finger on each hand. An extraordinary feature! These striking forelimbs terminated in a pointed claw that Alvarezsauridae may have used for ripping open termite mounds or for something else we haven't yet considered. In addition to its large eyes and monodactylous forelimbs, *Shuvuuia* possessed another rather extraordinary characteristic, which we'll discover a little further on.

CAMARASAURUS'S THIRD EYE

While many sauropods' semicircular canals are quite stunted, some members of this clade may have had another small asset to enhance their perception of their environment. Sauropod cranial scans often show a large sort of dorsal mass, which is sometimes connected to an opening or thinning of the bones at the top of the skull. Although the adjective "large" would never be used to describe a sauropod brain component, in any case, this structure is clearly visible, particularly in *Camarasaurus*. It may have been the home of a pineal eye, the "third eye" of some present-day reptiles and amphibians. Rather than an eye, it's a photosensitive organ that enables its owner to measure the intensity of ambient light. We can't rule out the possibility that some sauropods were equipped with this organ.

THE COLORS OF THE MESOZOIC ERA

To put it simply, many dinosaurs had large eyes and could see their surroundings well. What colors did they see? Answering this is difficult. We are trichromats (we have three types of cones in our retina), whereas most mammals and color-blind humans are dichromats. As for birds, they are tetrachromats, with a visible spectrum extending into the ultraviolet. The way they see the world is quite different from how we do. Less is known about the colors crocodilians see.[15] Some are thought to be trichromats (like the saltwater crocodile), and others tetrachromats (like the alligator). As for the dinosaurs, well, we don't know!

Whether they were di-, tri-, or tetrachromats, all the evidence suggests the colors of nature during the Mesozoic Era were likely not all that different from those of today, even though color diversity increased over time with the appearance of flowers, birds, and butterflies.

As for the dinosaurs themselves, they were just as colorful. Since the 2010s, paleontologists have been working to identify dinosaurs' colors by studying fossilized skin, which sometimes contains melanosomes detectable under the scanning electron microscope. These organelles, which are partly responsible for the color of bird feathers, have different shapes depending on the color they express: eumelanosomes are shaped like small rods and produce a black color, while pheomelanosomes are ovoid and produce a reddish-brown color. Thanks to the discovery of eumelanosomes, *Archaeopteryx* is now thought to have had black plumage[16] (and iridescent plumage like that of crows). The small carnivorous dinosaur

Sinosauropteryx that lived in what is now China had orange feathers and an orange-and-white ringed tail.[17]

In the absence of preserved melanosomes, the most ingenious researchers are still coming to bold conclusions. The presence of benzothiazole in its skin was therefore what led to the suggestion that the ankylosaur *Borealopelta* was probably red.[18] Benzothiazole is in fact a chemical compound of pheomelanin, and its presence is thought to be due to the degradation of pheomelanosomes. So far, we've identified red, white, black, orange, and reddish-brown in the dinosaurian color chart; at least, these are the colors produced by melanosomes. But present-day animals use other mechanisms to produce color, and there's little doubt we'll soon discover yellow, blue, and green in dinosaur skin.[19]

The Soprano and the Raptor

At the beginning of the 20th century, when nascent comic strips took up the interesting subject of dinosaurs, they often gave them cute little ears with rounded external auricles similar to mammalian ears. These cartoon dinosaur ears were often small, although a few cartoonists went wild, giving them rabbit or kangaroo ears. Sandor the atlantosaur (1911), whose creator is unfortunately unknown, and Maurice Cuvillier's formidable blue *Iguanodon* (1928) are examples featuring abnormally large external ears.[20] The phenomenon came to a halt in the 1930s, as cartoonists gradually came around to the general opinion that dinosaurs, like all reptiles, had no external ears, but simply a small hole toward the back of the skull. This small hole is, in fact, perfectly visible on the magnificent *Iguanodon* that the great popular science writer Camille Flammarion drew in 1886 for the poster announcing the publication of his remarkable book *Le monde avant la Création de l'Homme* (The world before the creation of man). The poster shows the *Iguanodon* reaching the sixth floor of a Parisian building and peering inside, which is a bit of an exaggeration. So how can three decades of big-eared dinosaurs be explained? It's probably a simple imitation effect between newspaper cartoonists, although it's not clear who started it.

Like olfaction, the abilities to hear and vocalize are unevenly distributed in the animal world. A soprano voice can emit a note of 1,048 hertz, while a bass voice can go down to 64 hertz. The frequency of human conversations tends to be between 100 and 300 hertz, with male voices being lower. Our auditory capacities are more extensive than our vocal possibilities, since the most sensitive among us can pick up sounds from 20 to 20,000 hertz, with sensitivity to higher frequencies decreasing with age. Bats express themselves in the ultrasonic range, from 15,000 to 150,000 hertz. On the other hand, some animals, from cassowaries to elephants, sometimes communicate using infrasound (below 20 hertz), enabling them to chat (inaudibly to humans) over long distances of up to 10 kilometers (over 6 miles)! Most birds have roughly the same range of auditory sensitivity as we do, which, incidentally, shows that ultrasonic bird and pest repellers are as inaudible to them as they are to us. The pest isn't always who we think it is.[21] Crocodilians, for their part, perceive low frequencies better than we do.

If we didn't have their inner ear to study, then the range of sounds dinosaurs heard might have been lost to us forever. Let's digress for a moment: Our external ears conduct sound to our eardrum, which then relies on the ossicles of the middle ear (in mammals) or the columella (in reptiles) to relay sound to the inner ear. The inner ear is a complex system of fluid-filled cavities, housed inside an osseous capsule known as the "bony labyrinth." Our labyrinth consists of the vestibule, the semicircular canals, and the cochlea. The cochlea is the part of the inner ear involved with hearing. Although dinosaurs didn't have auricles (or external ears, if you prefer), they did have a labyrinth with a cochlea. As the labyrinth is a cavity inside the skull bones, its shape and size can be determined by a CT scan. And the cochlea's proportions allow us to make several fascinating hypotheses about the audible frequency range for dinosaurs, based on research on hearing in birds.[22] For example, *Triceratops*'s frequency of best hearing would have been up to 290 hertz, but its high-frequency limit of hearing would have meant it only perceived sounds up to 1,500 hertz. In contrast, the tyrannosaurid *Gorgosaurus*'s frequency of best hearing went up to 1,500 hertz, with a high-frequency limit around 4,000 hertz. *T. rex*'s best hearing was up to a frequency of 580 hertz, and it could still hear up to its high-frequency limit at 2,100 hertz, while *Velociraptor* could hear perfectly up to a frequency of 2,400 hertz and still perceive sounds above 4,000 hertz. A soprano could have sung without alarming *Triceratops*, but unfortunately she would have drawn the attention of *Velociraptor*, who would have heard her perfectly.

Sauropods, on the other hand, were deaf. Their cochlea, like nodosaurids', was stunted. Their cousins, the ankylosaurids, however, had a long cochlea and a wider hearing range. Add to this the fact that they had a complex hyoid apparatus (which we'll discuss later) and nasal passages that could serve as resonance chambers, and you have reason to believe their vocalizations may have been far more elaborate than that of their cousins. As for *Triceratops*, not only was it insensitive to odors, it was also unable to detect frequencies above 1,500 hertz. On the other hand, it could hear low frequencies quite well and would therefore have easily perceived infrasound disseminating over long distances, just like elephants today. The lower the frequency, the longer the wavelength, which explains why only the bass from your neighbor's music reaches your home. Higher frequencies, with shorter wavelengths, are more easily blocked by obstacles. We don't know whether

Triceratops's frill could have served as an antenna to pick up sound waves, but we suspect it could have communicated by infrasound from several kilometers (more than a mile) away, which must have come in handy living in the same area as *Tyrannosaurus*.

In contrast, the hypertrophic cochlea of the curious little *Shuvuuia* is uncannily morphologically similar to that of the barn owl, whose exceptional hearing acuity enables it to hunt in complete darkness by orienting itself thanks to the sounds produced by its prey, and the sounds captured and amplified by its feathered facial disc. A similar lifestyle has been inferred for *Shuvuuia*, who would have been a formidable nocturnal predator.[23] Armed with large eyes and exceptional hearing, *Shuvuuia* was the master of the night in Mongolia's deserts during the Cretaceous Period.

Now that we have this extraordinary information, we can turn our attention to the soundscapes of the Mesozoic Era. In 1886, Camille Flammarion lamented, "Magnificent landscapes of bygone ages! No human eye has contemplated you, no ear has understood your harmonies."[24] Rest assured, a paleontologist has since looked into the matter![25] In the Triassic Period, the absence of birds, frogs, mammals, and the majority of insect families living today would have made for a relatively sinister soundscape for us. A few grasshoppers were already stridulating, however, and hemipterans (the order that includes cicadas) were already enlivening groves by "singing" with their tymbals. But no mosquitoes or bees were buzzing just yet. The various terrestrial vertebrates would have emitted sounds we can only imagine, such as jaw snapping, perhaps, in large amphibians, and hissing, gnashing, and jaw grinding in synapsids and the various non-dinosaurian archosaur groups. By the end of the Jurassic Period, when *Diplodocus* was roaming what is now the United States, the Alytidae family of primitive frogs was already croaking in chorus, with bazillions of orthopterans, hemipterans, and other dipterans adding their buzzes and stridulations. And then by the end of the Cretaceous Period, layered onto this hullabaloo must have been birdsong, hadrosaur and ankylosaur vocalizations, and ceratopsian infrasonic wails—in short, a terrible racket.

The Song of the Dinosaurs

At first glance, we might wonder if dinosaur song was even possible. After all, snakes and lizards are generally silent, preferring chemical communication. Turtles only make noises while mating, but the twenty-three present-day crocodilian species bellow, growl, and hiss.[26] Not to mention birds! In short, flanked phylogenetically by noisy animals, dinosaurs must have been chatty.

It's still difficult, however, to evaluate their voices. Did *Tyrannosaurus* roar with its mouth wide open like in all the movies? Or did it coo, mouth closed like the average pigeon? To answer this question, we need to look at some current data. The lion roars with its mouth open, thanks to an incompletely ossified hyoid bone that can vibrate. This is why your cat meows instead of roars—its hyoid bone is completely rigid. Meanwhile the crocodile grunts with its mouth closed at lower frequencies. In certain circumstances some bird species also vocalize with their beaks closed, but with one difference compared to crocodilians: crocodilians produce sound with the larynx, whereas pigeons emit sound from the syrinx, a cartilaginous area (sometimes ossified) located at the junction of the trachea and bronchi. All modern birds have a syrinx. Because crocodilians don't have this organ, they use their larynx to vocalize, just like you and me.

The syrinx is the vocal organ in birds, but it's not yet known when it appeared in evolution. We don't know whether dinosaurs produced sounds from their larynx like crocodilians or from a syrinx. It's interesting to note that in Greek mythology, Syrinx was a nymph who, to escape the advances of the god Pan, transformed herself into a reed. To console himself, Pan cut some reeds and glued them together with beeswax, making the first flute bearing his name, the panpipes, also called syrinx, from Greek.

Whether the sound is emitted by a syrinx or a larynx (which was not a mythological nymph, by the way), the pigeon and the crocodile are both able to express themselves without opening a beak or mouth. These closed-mouth vocalizations are particularly well adapted to territorial calls, as well as mating calls (such as "You have beautiful eyes, you know"). The esophagus or tracheal pouch dilates to enable the emission of these low-frequency periodic signals. In fact, the acoustic signal generated by the syrinx or larynx is strongly filtered when the mouth is closed, and only a short band of low-frequency acoustic energy is transmitted through the ventral part of the neck. Since birds and crocodilians use both types of vocalization (open mouth and closed mouth), is it a stretch to assume that dinosaurs did too? In birds, it's generally the largest species that vocalize with their mouths closed, so we could hypothesize that in dinosaurs too the emission of low-frequency closed-mouth sounds may have been linked to the size of various species, and used for territorial issues and other trifling matters. The fact that the dinosaur's inner ear was adapted to receive low-frequency sounds supports the idea that closed-mouth vocalization generates low frequencies. It would be a bit of a shame not to hear the sounds one makes!

As for the possible syrinx of non-avian dinosaurs, no trace of it has been found so far, which doesn't prove anything. The absence of proof is not proof of absence, especially when it's a cartilaginous organ and therefore unlikely to fossilize. Larynx or syrinx aside, many dinosaurs may have had an esophageal or tracheal pouch that dilated to produce closed-mouth vocalizations, whose modulation and tone, alas, remain a mystery. It's likely that *Tyrannosaurus* could vocalize with its mouth closed by inflating its esophagus pouch,[27] and perhaps it sounded a bit like a crocodilian bellow, but at this stage, we can't rule out the possibility that *T. rex* could have cooed, gobbled, or chirped.

Parasaurolophus, a North American hadrosaur with a gigantic tube projecting back from its skull, called a "cranial crest," is the first dinosaur whose voice has been able to be reconstructed. As with all its lambeosaurine cousins, the *Parasaurolophus*'s crest is a hollow tube, connecting the nostrils to the larynx via a long path along the tube, with exhaled air taking the same route. By replicating this crest and blowing air into it, American paleontologist David Weishampel was able to hear for the first time in seventy million years the gentle voice of a dinosaur—a modulated, low-frequency sound that might recall the timbre of a bassoon or tuba.[28] By scanning the skulls of other lambeosaurines, such as *Corythosaurus*, *Lambeosaurus*, and *Hypacrosaurus*, other North American researchers have shown that hadrosaur cranial crests all had a structure comparable to that of *Parasaurolophus*, with long cavities for the passage of air.[29] Since resonant frequency of a tubular structure is related to its length, the longer it is, the lower the frequency of the sounds produced. This also led to the reasoning that juveniles, whose crests are not fully grown, would have produced much higher-frequency sounds than adults. By measuring the cochlea's length, it was possible to confirm this exciting hypothesis of vocal communication in hadrosaurs: the best frequency of hearing was around 580 hertz in the smallest specimen (a young *Hypacrosaurus*), compared with 80 hertz in the largest (an adult), with corresponding maximum high-frequency hearing limits at around 2,100 hertz and 1,200 hertz, respectively. The adult could definitely hear a high enough frequency to pick up the squeaks of its offspring.

Ankylosaurs also had convoluted nasal passages but without the elaborate cranial crest of the *Lambeosaurus*. Once again, X-ray microtomography was used to reconstruct *Euoplocephalus*'s complex nasal cavities, which form loops behind the nostrils. This bizarre structure could have served as a resonance chamber during vocalization, as in the case of *Lambeosaurus*.

Until more data on dinosaur vocalizations becomes available, we'll remain cautious about characterizing them more precisely, such as how we say that "the camel grunts, and the owl hoots."[30] We know that the magpie chatters and the jay whispers. But hadrosaurs? Or tyrannosaurs? If hadrosaurs were perhaps buzzing like a bassoon, what words could be used to describe the song of other dinosaurs? In the absence of a scientific hypothesis, let's leave the answers to the novelists for the moment; in *Les semeurs d'épouvante* (Sowers of terror), a fantastic "novel of Jurassic times" published in 1923, Fernand Mysor bravely tackled the question. He decided that plesiosaurs meowed, labyrinthodonts croaked, *Diplodocus* roared, stegosaurs and ichthyosaurs bleated, pterodactyls snarled, *Compsognathus* barked, and *Iguanodon* growled.

Ampelosaurus Couldn't Say No

Ampelosaurus, the "vineyard lizard," was discovered in the Upper Aude Valley in southern France. Among this dinosaur's many bones were half a dozen from the back of the skull, the bony case containing the brain and inner ear. In 2020, we decided with a brilliant master's student, Gwendal Adam, to find out more about this dinosaur's intellectual abilities and sense of balance. When we used X-ray microtomography to study the skull, with the help of University of Montpellier researchers Renaud Lebrun and Lionel Hautier, we didn't expect to find signs of prodigious brain activity. We weren't sure we'd find anything, as *Ampelosaurus* bones are highly mineralized. But after a few dozen hours of image processing, the miracle of tomography revealed the unknown cavities of the skull to us:

brain, cranial nerves, and semicircular canals of the inner ear. This is how a fossil whose scientific interest was thought to have been exhausted suddenly reveals a wealth of new information. Nevertheless, discovering that *Ampelosaurus* had such a small lateral semicircular canal almost saddened us, as lateral head movements must have been almost nonexistent in this dinosaur. The anterior semicircular canal was slightly larger than the posterior semicircular canal, suggesting that, while *Ampelosaurus* couldn't shake its head laterally, it could properly nod its head. In any case, this dinosaur had a wider range of head movements in the sagittal plane, probably related to its feeding habits.

Tyrannosaur Kisses

Crocodilians have integumentary sensory organs (ISOs) in some of their scales that detect temperature, touch, and chemical compounds. ISOs are therefore thermoreceptors, mechanoreceptors, and chemoreceptors, and are particularly abundant in crocodilians' snouts, where the bones are perforated by multiple small orifices that allow hundreds of branches of the trigeminal nerve to pass through. The trigeminal nerve transmits high-resolution information to the brain. Crocodilians' ISOs are highly sensitive, making their snouts more sensitive to touch than primate fingertips.[31] And of course, the snout of the tyrannosaurid *Daspletosaurus* had exactly the same foramen as the snout of crocodilians,[32] demonstrating that some dinosaurs also possessed ISOs and must have been able to use them like crocodilians. Tyrannosaurs must therefore have had a highly sensitive facial tactile system, which they used to capture prey, identify and manipulate objects, and sense temperature. Female tyrannosaurids could have used the ISOs on their snout to maintain optimum nest temperature as well as to gently grasp their eggs or babies between their teeth to move them, just as female crocodilians do. And during courtship rituals, tyrannosaurs might have rubbed noses to get to know each other better before getting down to business.

Crocodilians also have ISOs scattered all over their bodies; their skin is therefore both thick and protective, and highly sensitive thanks to these sensory receptors. The presence of scales quite morphologically similar to crocodilians' on the tail of the small carnivorous dinosaur *Juravenator*, which frequented the lagoons in what is now Germany during the Jurassic Period (this area was packed with tropical lagoons, an ideal vacation destination in the Late Jurassic), led two researchers to postulate that it also possessed ISOs along its tail.[33] They speculate that, like modern crocodiles, this Jurassic lagoon-dweller could fish by night, picking up the slightest ripple and molecule floating in the water to orient itself. This is a credible hypothesis, since fish remains were discovered in the preserved stomach contents of some of *Juravenator*'s cousins.

The recent discovery of these sensory organs in dinosaurs, along with a better understanding of their function in crocodilians, opens up new perspectives in our quest to understand how dinosaurs perceived their environment. In addition to their ears, eyes, and nose, at least some dinosaurs (and probably most to varying degrees) were equipped with other effective sensory organs. Thanks to their ISOs, they simultaneously had a temperature sensor, chemical sensors, and a sense of touch more sophisticated than our own.

Dinosaur Nociception

As carnivorous humanists, we'd prefer for animals not to feel pain, or more precisely, not to feel anything, especially the ones we eat. Unfortunately for us, this is not really the case. Not only do mammals feel pain, but so do reptiles, birds, and probably even fish. Palliative-care experts have recently looked into traces of dinosaur nociception (the sensation of pain), recalling that the evolutionary function of pain is to incite an injured animal to protect its wound in order to heal.[34] The experts' conclusion was that the lives of these poor animals, who didn't have access to skilled veterinarians, was not a path strewn with roses. Healed fractures of the fibula, ribs, and toes were numerous. We even know of a sauropod from the area that is now Thailand with a poorly healed humerus following a fracture. The bone broke in two, and then the two halves turned and rejoined that way. The crippled animal probably finished its life with three functioning legs, but years after its accident! A humerus from the hadrosaur *Hypacrosaurus*, from present-day Canada, tells much the same story. After the bone was broken in two by an oblique fracture, the pieces spread apart, rotated, and then healed in incorrect alignment. Barnum Brown, who found the bone, said it was the sickest bone he'd ever seen,[35] and he had seen many other bones. Fatigue fractures (or stress fractures) are also frequently seen in the finger bones. These small cracks in the bone are caused by repeated, intense movements, such as from walking. Some strange small depressions have also been discovered on tyrannosaur phalanges, evidently bone erosion caused by gout. Gout, a disease that causes painful inflammation in and around the joints, must have already been causing a lot of pain in the Cretaceous Period. Since a diet based on vegetable proteins is strongly recommended to combat gout attacks, *Tyrannosaurus* was at high risk! Maybe this is the reason he was a bit grumpy.

Dinosaur tracks have also shown evidence of misery.[36] In Morocco, a theropod trackway from the Middle Jurassic Period was discovered that betrays a pathology: an irregular gait. The stride length of alternate steps differed, and the two outer toes of the right foot were almost stuck together. In lay terms, the poor dinosaur limped and had a deformed right foot. Other theropod footprints have been found that also show deformed digits, perhaps due to swelling, or partial or full amputation. Theropods with limps, others missing a toe—it was a world full of hardship! It's worth noting that fractures of weight-bearing bones (tibiae and femurs) were fatal in bipeds, but they were able to recover from just about everything else. A hadrosaur from Canada, for example, recovered from a bilateral fracture of the ischia (lower part of the pelvis), "the most serious example of hadrosaur osteopathy," according to paleontologist Darren Tanke. It is indeed surprising that the animal survived such a traumatic injury, but dinosaurs were undoubtedly tough cookies. This type of near-miraculous recovery is a plausible sign that the wounded dinosaur was cared for by its fellow dinosaurs while it healed.

A team of paleontologists and oncologists recently identified osteosarcoma, a highly aggressive bone cancer in humans, on the fibula of a *Centrosaurus* (a cousin of *Triceratops*).[37] An osteosarcoma this advanced and untreated would be fatal to a human patient, the researchers noted, but it was not the direct cause of this ceratopsian's death. What's more likely is that it drowned crossing a river with its friends, as we'll see later.

Without a doubt, the winner in the "slow and painful death" category was a tyrannosaur nicknamed Sue. The skeleton preserved at Chicago's Field Museum displays an impressive catalog of traumatic lesions. On the right side, there are three broken (and healed) ribs, as well as damage to the scapula and humerus, likely the consequence of a violent impact from which the animal recovered. On the left side, five ribs were broken and healed. Sue's left fibula also shows bony outgrowths, perhaps the aftereffects of a fracture. Worse still, and probably lethal, perforations in Sue's lower jaw have been interpreted as evidence of infection by a protozoan parasite of the *Trichomonas* genus, which is known to infect birds. Sue was suffering from advanced trichomoniasis: since the jawbones were perforated, numerous oropharyngeal lesions would have been present, as well as necrotic ulceration of the mouth and esophagus.[38] Feeding must have been very difficult at this stage of the disease, and it's likely that Sue died of starvation. As if starving weren't the ultimate punishment for such a glutton, a Canadian researcher discovered three deformed teeth in Sue's jaw, two of them fused together—another consequence, according to the researcher, of a *Trichomonas* infection. This possibly resulted in a horrendous toothache to add to the final moments of Sue's agony.[39]

Finally, in the "I've survived horrible injuries" category, a hadrosaur wins again. The skull of a poor *Edmontosaurus* was discovered in South Dakota, and this fossil "mummy" had quite a lot of skin preserved on the skull bones.[40] A scar is visible in the middle of this skin. The scales had grown back around the wound with a different arrangement from the rest of the skin: the healed wound is smooth, and new scales have grown all around it, forming a structure with small radial wrinkles. This is similar to how wounds heal in modern-day lizards. We are therefore in the presence of the first discovered dinosaur paleo-scar. The skull discovered with the preserved skin is even more telling as it bears large teeth marks on top, and the bones of the right orbit have been torn out. These traumatic injuries all show signs of reossification, which testifies to the victim's survival of this violent trauma. In short, this unlucky dinosaur had its right eye ripped out along with part of its face but went on with its life, one-eyed and disfigured.

The existence of these severely injured dinosaurs that recovered suggests high healing capabilities, as well as the possibility of mutual aid during convalescence. An injured dinosaur may have been protected or even fed by members of its group. Dinosaurs also exhibit the reactions expected of animals sensitive to pain, such as the limping that can be seen in their footprints. Since these creatures were undoubtedly sensitive to pain, perhaps their extinction provides some consolation—they'll never have another toothache.

Argentinosaurus's Siesta

The life of a dinosaur was not an easy one. A little rest now and then couldn't have hurt. But when it comes to the quality and quantity of dinosaur sleep, there are still many mysteries. While REM sleep has been known to exist in mammals and birds for a long time, researchers believe they have observed it in recent studies of lizards, strongly suggesting it may have been present in dinosaurs as well.[41] But it may be hard to find the proof! At first glance, another unanswerable question concerns the sleep duration of dinosaurs, especially large ones. We still need to agree on what sleep is in different vertebrate groups. Researchers have shown that the bigger a mammal is, the less it sleeps. Elephants sleep for just two hours a night—true insomniacs![42] They devote an enormous amount of time to eating, from sixteen to twenty hours a day, and that doesn't leave much time for snoozing. Consuming vegetation at a rate of 10 kilograms (22 pounds) per hour, it takes elephants all day and most of the night to feed. So thinking of the sauropod *Argentinosaurus*, one of the largest known land animals of all time that weighed 70 metric tons and measured 35 meters (almost 115 feet) long and 8 meters (more than 26 feet) at the withers, it probably had to swallow hundreds of kilos of vegetation every day. What's more, its days were shorter! Every century, the length of a day increases by 1.7 milliseconds. Not much admittedly, but *Argentinosaurus* lived 95 million years ago, the equivalent of 950,000 centuries, so the day has lengthened by 1,615 seconds, or almost 27 minutes, since then. So did the colossal *Argentinosaurus* sleep, and if so for how long? Using the regression between the sleep duration and mass of a herbivorous mammal published by the South African researchers who studied sleep in elephants, *Argentinosaurus* would have slept less than an hour a day (that said, it was at least something given that the days were shorter, and the nights too). Of course, extrapolating conclusions about a dinosaur from mammalian data is questionable, to say the least, but it's better than nothing.

ASLEEP ON THEIR FEET

Since we're on the subject of intriguing questions, here's another: did *Argentinosaurus* sleep standing up or lying down? Elephants in the wild usually sleep standing up, but they sometimes lie down too. We don't know if an *Argentinosaurus* could lie down. It's doubtful because, even more than lying down, the problem would be raising its 50 metric tons to stand back up without the help of a crane. On the other hand, several resting places of much smaller dinosaurs have been discovered. Scholars of paleoichnology have poetically named the traces left by animals resting in the mud "cubichnia" (to distinguish them from "repichnia," which are traces of movement). Many of these traces were from the Lower Jurassic of Connecticut, where Reverend Hitchcock (1793–1864), the reluctant founder of dinosaur paleoichnology, discovered and described hundreds of fossilized footprints. We can call him "reluctant" because Hitchcock thought he had discovered bird tracks, and he never admitted they could be dinosaur footprints. But the resting traces of various theropods and ornithopods show the impression of metatarsals, the ends of the ischium and pubis, and sometimes the hands too. Therefore, these bipedal dinosaurs squatted like ostriches or emus.

NIGHTTIME IN CHINA

And sometimes a little snooze turned into eternal sleep. *Mei long*, meaning "sleeping dragon" in Chinese, was a troodontid from China measuring 53 centimeters (21 inches) long. Its name refers to the position in which it was fossilized, curled up on itself with a wing (or feathered forelimb, if you prefer) covering its head—in other words, the sleeping position of many modern birds. The animal must have fallen asleep quietly one night, never to wake up again.[43]

This natural need to get some sleep gave researchers an idea for what *Tyrannosaurus*'s much-talked-about atrophied little arms were for. *Tyrannosaurus* certainly led a tiring life devouring hadrosaurs and ceratopsians from breakfast to dinner, so it must have enjoyed a well-deserved rest from time to time, and to do that perhaps it squatted like its distant great uncles in Connecticut. When it was time to wake up—and the world undoubtedly belonged to the early rising *T. rexes*—what better way to guide the lifting up of its 7 metric tons than those two little arms, like a sprinter off to a flying start? This is only a hypothesis, and it requires confirmation, something paleoichnology could certainly provide.

A LITTLE PHYSIOLOGY

It should be noted, however, that there is still some doubt as to whether reptiles experience REM sleep, and it has been suggested that only homeothermic species do. A little vocabulary to clarify things: A homeotherm has a constant body temperature no matter the environmental temperature (and is also called "warm-blooded"), unlike a poikilotherm, whose internal temperature fluctuates according to the environmental temperature (a.k.a. "cold-blooded"). Similarly, an endotherm has a metabolism that enables it to maintain its internal temperature, unlike an ectotherm, whose body heat comes from the environment (usually the sun). In short, humans are homeothermic endotherms, like birds, but crocodilians are poikilothermic ectotherms, which gives them an excellent excuse to nap in the sun for half their lives. And what about dinosaurs? Were they warm-blooded or cold-blooded?

These questions of homeothermy/poikilothermy in dinosaurs have been resolved in recent years by geochemists, individuals

I should warn you that we should be terribly wary of. They'll ask sweetly if they can take a few milligrams of the one and only *Ampelosaurus* tooth that we have, which itself weighs just a few grams—for the greater glory of science, of course. Now it is true that geochemistry has made great strides in recent years, and this would only be a sacrifice of a few milligrams . . . In short, while it's difficult to be a geochemist and still be on good terms with museum curators, it does happen. And geochemists have settled a decades-old debate on dinosaur metabolism by examining the carbon and oxygen isotopes in dinosaur teeth and eggs. The relative abundance of the different isotopes reflects the body temperature at the time of tissue formation. By measuring the proportions of oxygen-18 and oxygen-16 in a sample, we can deduce the dinosaur's internal temperature—an excellent paleothermometer![44] The body temperature of sauropods is now estimated to have been between 35 and 38 degrees Celsius (between 95 and 100.4 degrees Fahrenheit), and that of oviraptorids around 32 degrees Celsius (89.6 degrees Fahrenheit). A healthy hadrosaur *Maiasaura* would have had an internal temperature of 44 degrees Celsius (111.2 degrees Fahrenheit), a temperature found in some modern birds.[45] To sum it up, we can say that dinosaurs were generally warm-blooded animals, capable of maintaining a high internal temperature no matter their environment's temperature. So, they were homeothermic endotherms like birds and mammals.

4

Mesozoic Sociology

After our look at the intellectual and sensory capacities of several dinosaur species, we still need to understand what they did with these capacities in their daily lives. Were they fierce, solitary creatures defending their territory against encroaching congeners, or did they gather in colonies like penguins on their ice floe? In nature today, there are a few species of asocial vertebrates who only cross paths with others to reproduce, but most species are relatively social, whether only at certain times in their lives or on a permanent basis. Before venturing into the details of how "paleopacks" worked, it's important to be aware of the fact that we only know this kind of thing about a tiny fraction of species living today: knowing whether giraffes live in stable groups or flit from one group to another, or whether elephants sleep standing up or lying down requires hundreds or thousands of hours of observation, and we don't have enough ethologists to know everything about the behavior of all animals! Reptiles are among the least studied, no doubt because they don't meet the cuteness criteria. Compared to a bird or a small mammal, a crocodile, even a tiny one, just isn't that cute. And as their economic value is limited (although crocodile handbags are very expensive), behavioral studies of present-day reptiles are still rare.[1]

As for their extinct dinosaur cousins, is it possible for us to have at least some idea about their social behavior and interactions? Surprising as it may seem, the answer is yes, based on two criteria: footprints (ichnology) and paleontological evidence of mass mortality. The second criteria is not always relevant, because the accumulation of bones in a given place does not necessarily reflect the composition of an ecosystem, let alone that of a population. Most paleontological deposits were formed by the accumulation of carcasses or bones over relatively long periods. Remains carried away by water are deposited in a specific place for topographical and hydrological reasons over a span of months or decades. For example, remains may accumulate in the bend of a river, and it would clearly be inaccurate to conclude that the animals whose remains were found there lived together. Occasionally, however, certain fossil sites attract attention because they are composed entirely of individuals of the same species, and even of articulated skeletons stacked one on top of the other. In these cases, we consider the hypothesis of a particular event having caused mass mortality in the same population, as in the present-day example of wildebeests drowning by the hundreds while crossing rivers during their annual mass migration, much to the delight of the crocodiles their carcasses feed. As for the first criteria—ichnology—it is certainly more relevant to assessing sociality: discovering parallel tracks from animals of the same species moving together in the same direction and at the same speed is a strong argument for the hypothesis of herd life in this species. By combining these two types of information, we can try to discern some key features in the social behavior of large groups of dinosaurs.

Antisocial Dinosaurs

The skeletons of dinosaurs that lived solitary lives have typically been found in isolation. No mass mortalities, no tracks providing evidence of herds. But who were these irritable, awkward hermits? It's a bit by default that we attribute this lifestyle to stegosaurs, munching their ferns all by themselves. It's true that most stegosaur genera have only been identified by a few bones or a single partial skeleton, and this scarcity remains the main clue indicating their supposed agoraphobia. This therefore led to the assumption that *Stegosaurus* walked its 9-meter (29.5-foot), 6-metric-ton body around all alone, only meeting up with its fellow *Stegosaurus* from time to time to ensure the survival of the species. Given the morphology of these animals, these amorous encounters were high risk, a topic we'll return to later. There is, however, one exception to stegosaurs' apparently antisocial behavior: German paleontologists on an expedition in Tanzania just before the First World War discovered a deposit containing the disarticulated remains of at least thirty *Kentrosaurus*,[2] which may make us skeptical of our assumptions about the animal's habits. In fact, it's highly probable that *Kentrosaurus* was a social animal, but we're still not sure about its cousins.

It's a similar story for adult ankylosaurs, which are generally very rare to find in paleontological sites and are considered fans of social distancing. *Ankylosaurus* fossils account for just 0.05% of the numerous dinosaur specimens discovered in the Hell Creek Formation in the United States. There is no trace of sociality in this enormous herbivore that measured 8 to 10 meters (26.25 to 32.8 feet) in length and weighed between 6 and 8 metric tons;[3] it would have been an antisocial loner. Thyreophorans as a group (which includes stegosaurs and ankylosaurs) may well have been a bit gruff, if not downright sociopathic, and even slightly paranoid if you consider ankylosaurs' bony full-body armor as part of their clinical presentation. But, as we'll see further on, this apparent desire for solitude developed later in life for them.

Lifelong Groups

Apart from our dour thyreophorans, most dinosaurs seem to have been highly social. Sites where footprints have been discovered often show carnivorous dinosaurs of various sizes, with their characteristic tridactyl footprints, taking a walk together. Were they chatting about the weather? Going hunting? We'll come back to this later, but in any case, both small and large theropods were happy to stroll around together.

This was also the case for the gigantic sauropods, which were long thought to have spent their lives lazing around in lakes, far too heavy to frolic on dry land. This image, which many of us have stored somewhere in the back of our minds, was shattered nearly a century ago—which says a lot about the quality of the brain's updates! But let's take a closer look at the circumstances surrounding the sauropods' "emergence from the water," which is owed to the efforts of an American researcher named Roland Thaxter Bird. Nicknamed R. T. by his friends and family (pronounced "Ertie"), he spent a decade as one of the fossil collectors for the legendary Barnum Brown of New York's American Museum of Natural History. I invite you to read more about Brown on your own, as sadly I can't devote several chapters to him in each of my books.[4] Bird was employed by the museum's paleontology laboratory to collect fossils across the United States according to his boss's whims. He had spent the summer of 1938 in Montana without much success, only to end up in New Mexico in November based on vague instructions sent by Brown. Bird had the good sense to write his memoir,[5] so we can follow his travels through the United States. His quest began somewhat by chance, when a young man became interested in the large fossils Bird was transporting. Having big dinosaur bones in your trunk is a great conversation starter. In fact, I could tell you about a long discussion I had with Swiss customs officers one day when they had me open my trunk at the border, but I digress. Like most people who come across a paleontologist, the young man couldn't resist recounting his own experience of paleontology. He told Bird where to find fossilized footprints of 12-foot giants in a Native American shop in Gallup. Looking up at the sky, Bird saw it was getting late, and since it was also windy and snowing pretty hard, well, he thought spending the night in Gallup might be the best option. And in that case, why not take a little detour to this shop? There, interesting footprints awaited him in a display case: perfectly recognizable human footprints, 50 centimeters (almost 20 inches) long and carved out of stone by human hands. After speaking with the salesperson, Bird learned that more tracks, including dinosaur tracks, were on display at another shop in Lupton, Arizona, about 23 miles from Gallup, so he sped off in his old Buick. In Lupton, the plot thickened: next to more carefully sculpted human footprints were perfect tridactyl theropod footprints—too perfect, in fact, and clearly the work of the same skilled stonecutter. Bird then learned that all these fake footprints came from Glen Rose, Texas, about 800 miles away. Intrigued by the quality of the fake dinosaur footprints, he decided to go take a look in case they had been copied from real ones. And the trip did not disappoint him. All around Glen Rose, dinosaur footprints were everywhere, from the walls of houses to the Paluxy riverbed.

Bird found several tridactyl footprints of theropod dinosaurs in the Paluxy riverbed, but nothing out of the ordinary. This type of track had been known for a good century. Following Barnum Brown's wise motto ("Always dig three feet beyond"), in a last-ditch effort, Bird finally uncovered a large depression with claw marks, then a second, and a third. It was the first discovered sauropod trackway—or at least the first to be recognized as sauropods'. When he brought a plaster cast of these footprints back to New York, Bird was greeted by a doubtful Brown, who eventually said the casts were all very well but that Bird would have to bring back the real thing. So the enormous slabs of rock were excavated from the Texan riverbed and brought to the second floor of the American Museum of Natural History. This may seem like a harrowing, fanciful, and even impractical operation, but for Brown and Bird, it was an absolute dream, the perfect project for the short digging season. Bird considered the discovery significant enough (and it was) to propose a short article on the subject to the journal *Natural History* to popularize the find. In his article, he proudly announced the first discovery of sauropod tracks and the possibility that these creatures were land animals. But he also revisited the colorful little detour of his trip, where he had been shown 20-inch human footprints laboriously carved into rock. By reporting on this unusual handicraft, Bird had no idea he would unleash the passions of Young Earth creationists. According to them, Earth is around six thousand years old, and humans and dinosaurs coexisted. Many people who read Bird's article found material in it to support these fantasies of human-dinosaur coexistence, something Bird complained about in his older years, outraged at having been quoted in a bunch of perfectly ridiculous creationist publications.

In the summer of 1940, Bird set off again to carry out his beloved boss's wishes. Meanwhile in Bandera, Texas, young girls discovered "elephant footprints," which caught the attention of Texan geologists, and the information inevitably got back to Barnum Brown in New York. Brown promptly sent Bird south in search of sauropod footprints. At Davenport Ranch, near San Antonio, Bird and his team uncovered twenty-three parallel trackways that showed juveniles and adults moving together in the same direction. But none were suitable to take to the museum in New York, and in any case, the ranch owner objected to any of them being removed. Bird had no choice but to return to Glen Rose. On site, he and his team began building a cofferdam with sandbags in the middle of the

Paluxy River, then they drained the area. When excavating a riverbed, every storm is a curse, as the dikes quickly become submerged. That summer was particularly rainy, and the exercise of emptying the excavation area had to be repeated half a dozen times. But by the end of the summer, hundreds of footprints had been uncovered, including numerous sauropod trackways. The trackway Bird decided to bring all the way back to New York was slightly curved and, the icing on the cake, included a carnivorous dinosaur trackway that curved in the same direction. Bird and other paleontologists after him defended the hypothesis that these tracks showed this theropod was pursuing the sauropod. Looking at the site as a whole, the two trackways in question are not isolated: a dozen parallel sauropod trackways have been uncovered at this site since 1940, as well as three trackways of large carnivores. The most plausible interpretation is that, after a sauropod tribe passed through, three *Acrocanthosaurus* passed by the same spot. There's nothing to indicate that they were hunting herbivores, or indeed that they weren't. Finally, at the end of 1940, almost 9 meters (29 feet) of sauropod trackway—about 40 metric tons of rock—were sent to the American Museum of Natural History, according to Brown's wishes. While Bird did not discover clear proof of sauropod hunting, he was the first to demonstrate the existence of sauropod herds, as well as the fact that these animals were terrestrial and not amphibious. Incidentally, this 9-meter trackway from the Paluxy River is still on display at the American Museum of Natural History beneath an *Apatosaurus* skeleton in the configuration Brown had in mind back in 1938. A slight anachronism, seeing as the skeleton is around 150 million years old and the trackway is more like 105 to 110, but we're happy to forgive the paleontologist. For all he contributed, I can't help but give that man a free pass.

Imagine a pack of sauropods advancing majestically in concert—it must have been a breathtaking sight! Maybe the ground was even shaking, like in *Jurassic Park*! It was, however, a confoundingly commonplace sight in the Jurassic and Cretaceous Periods. Since Bird's first discovery, sauropod trackways have been found all over the world, confirming that these animals readily traveled in herds. Numerous footprint deposits show parallel trackways of several large dinosaurs, sometimes more than twenty and even as many as forty.

The Cretaceous Wildebeest

In *Centrosaurus*, a ceratopsian from the Campanian Stage of Canada, downright delusional herd behavior was a problem. Many paleontological deposits in the province of Alberta contain mainly *Centrosaurus* bones, sometimes in the thousands. These various bonebeds (known as the Hilda mega-bonebed) were all found in the same stratum and may have been formed simultaneously by a gigantic flood that drowned thousands of animals from a single herd (the migration hypothesis).[6] This scenario occurs every year in the savannas of East Africa when huge herds of blue wildebeest (*Connochaetes taurinus*) cross swollen rivers. Migrating wildebeest herds can number in the tens of thousands, leaving hundreds or thousands of corpses behind to the delight of crocodiles. It's believed that, during the Cretaceous Period, the ceratopsians forming the large single-species bonebeds also lived in huge herds on the plains of what is now Canada, where they had to travel great (or not so great) distances. In addition to *Centrosaurus*, these gregarious species included its cousins *Styracosaurus* and *Pachyrhinosaurus*, but not the iconic *Triceratops*, for which no mass mortalities have yet been confirmed. The remains of a few *Triceratops* have sometimes been discovered at the same site, but nothing comparable to the hundreds or thousands of *Centrosaurus* that died together in Alberta (one of which, remember, was afflicted with a nasty bone cancer). It is plausible that *Triceratops* lived in small family groups.[7]

Returning to the blue wildebeest, some of the many studies devoted to them focus on the geometry of the "front patterns" of their immense herds, which extend over several kilometers (more than a mile).[8] The front pattern of a herd is considered to be an example of swarm intelligence. This concept, of particular interest to computer scientists, postulates that the combined efforts of nonintelligent creatures produce collective intelligent behavior. The wavy front of a wildebeest herd shows signs of swarm intelligence; without discussion or consensus, the animals systematically explore and overcome all obstacles. What about the wavy fronts of *Centrosaurus* herds? The discovery of mega-bonebeds suggests that swarm intelligence—in centrosaurs as in wildebeests—didn't always work.

The small ornithopod *Dysalotosaurus* lived in East Africa at the end of the Jurassic Period. It too liked to travel in numbers and drown in crowds in riverbeds. At least, that's the modern interpretation of a Tanzanian deposit excavated between 1909 and 1913 by German paleontologists. Tanzania was a German colony at the time, and paleontologists from Berlin spent several years carrying out excavations of colossal proportions—probably the biggest ever—using hundreds of native workers to excavate all around Tendaguru Hill. The most emblematic of their discoveries is the immense skeleton of *Giraffatitan*, a beautiful animal whose head reaches a height of 13 meters (over 42 feet) and that can be visited at Berlin's Natural History Museum. But many other less spectacular dinosaurs were also discovered by German researchers. One of the quarries to the northeast of the hill yielded fourteen thousand bones, several thousand of which were from *Dysalotosaurus*, one of the world's most widely documented dinosaurs and unjustly little known. For instance, had you ever heard of *Dysalotosaurus* before now? *Dysalotosaurus lettowvorbecki* is named to honor Paul von Lettow-Vorbeck, a successful German general in the First World War. He operated in the African Great Lakes region with native troops against the British and their African troops. The fact that he was never defeated in battle and successfully invaded British-administered territories, claiming thousands of lives, earned him great renown in defeated Germany and led a dinosaur to be named after him in 1920. *Dysalotosaurus* (the "uncatchable lizard," like the general) is a small ornithopod dinosaur of the Dryosauridae family and a small cousin of *Iguanodon*. They grew as large as 2.5 meters (8.2 feet) long, and it appears from the Tendaguru findings that they lived in herds of mixed ages, unlike many dinosaur species. *Dysalotosaurus* lived in extended families, ranging in age from juveniles newly hatched from their eggs to wrinkly, old individuals. The skull differences seen between the youngest and the adults also suggest that, while the latter must have been strictly herbivorous, the young may have had a more varied diet.

Youth Gangs

Once you've shared the same nest, what could be better than spending your youth together? For dinosaurs, childhood was often brutally interrupted by some predator: carnivorous dinosaurs, of course, but also crocodiles, lizards, snakes, turtles, and even mammals! While our furry cousins were relatively few in number in the Mesozoic Era, some grew to be fairly large, like *Repenomamus*, whose size rivaled that of a badger and who roamed China 125 million years ago.[9] And this badger-sized animal (it wasn't a badger at all, by the way, but a gobiconodontid) could think of nothing better than gorging itself on baby *Psittacosaurus*, whose remains have been found preserved in a *Repenomamus*'s stomach. In short, to increase the chances of surviving attacks from hungry mammals and other voracious reptiles, or even from one's own aunts and uncles (cases of cannibalism have been reported in *Coelophysis*), staying close with other juveniles was a good choice. In the event of an attack, a few members of a group can survive, while an isolated animal would be killed. Nevertheless, survival rates must have been pretty low in the Mesozoic jungle! Several dinosaur species were partial to forming youth groups, including the ankylosaur *Pinacosaurus* and the marginocephalian *Psittacosaurus*, as well as sauropods and theropods.

Accumulations of young *Psittacosaurus* skeletons are common, and often up to thirty are found together. The most remarkable example dates to the Lower Cretaceous of Liaoning, a province in northeast China. At least twenty-four tiny *Psittacosaurus* specimen were found fossilized in the same place (their skulls measured 3 to 5 centimeters [1.2 to 2 inches] in length, and their femurs 3 to 4 centimeters [1.2 to 1.6 inches]), along with the skull of a larger individual (this skull measured 12 centimeters [4.7 inches] long). These *Psittacosaurus* were not fetuses, however, but juveniles that had been hatched for several months. And the "big" skull wasn't so big either, since its owner's age has been estimated at four to five years old, well below sexual maturity—which would have occurred at around eight to nine years in *Psittacosaurus* (methods for determining the age of dinosaurs and the moment of their sexual maturity will be discussed later). So this wasn't a parent-child group, but an adolescent-child one. Another group of six *Psittacosaurus* juveniles was found in China, probably killed by a lahar (a mudflow of volcanic debris). They weighed between 300 grams and a kilogram (0.7 to 2.2 pounds), with femurs measuring 5 to 7 centimeters (2 to 2.75 inches) long. Their estimated age at death was between eighteen and thirty-six months, indicating that these animals came from different clutches. And there's every reason to believe these groups of fossilized animals led their lives together.[10]

Among *Psittacosaurus*, groups of young seem to have been entrusted to a big brother or sister, or an uncle or aunt, but unfortunately we still can't tell the boys from the girls, or the uncles from the aunts. The intellectual performance of these fairly primitive marginocephalians was rather superior to that of their distant successors, the ceratopsids, and they may have had fairly complex societies, with adolescents possibly responsible for looking after babies and young individuals. These groups of newly hatched young are reminiscent of present-day crocodilian nurseries, where freshly hatched young from several different clutches gather under the supervision of a few adults for several months.

Protoceratops is another small marginocephalian that lived in Asia at the end of the Cretaceous Period, particularly in Mongolia, where numerous skeletons have been found. According to historian Adrienne Mayor, it's possible that ancient nomads found fossilized remains of *Protoceratops* while traveling through this part of the world several millennia ago and that discovery gave rise to the legendary griffin, an animal with the head and wings of an eagle and the body of a lion.[11] It's certainly far-fetched, although *Protoceratops* is indeed a quadruped with a beak, like the griffin. But one thing is certain: at every stage of their lives (newborn, juvenile, subadult, adult), *Protoceratops* lived in groups determined by age. Fossil accumulations of each of these age classes have been found with no intergenerational mixing.[12] For the *Protoceratops*, family wasn't mom and dad; it was the brothers and sisters and cousins with whom they were born, grew up, reproduced, and died. There was no mixing of the old with the young or the young with the old. It was a successful technique for avoiding generational conflict.

The ankylosaur *Pinacosaurus* also lived in what is now the Gobi Desert. Fossilized remains of numerous young *Pinacosaurus* have been found together in at least three deposits. They're frequently found standing, in the position in which they died, probably during sandstorms. We are therefore absolutely certain that they lived together. A dozen 2-meter-long (6.6-foot-long) animals were found in the Bayan Mandahu Formation, and over thirty of the same size in the Alagteeg Formation. Neither adults nor young individuals were found in these two famous Mongolian deposits—only adolescents! While their parents withdrew from the world for some peace and quiet, young ankylosaurs seem to have stuck together, living in age classes like many other dinosaur species.

Similar patterns are found in saurischians. The Suhongtu Formation in Inner Mongolia in the south of the Gobi Desert has yielded skeletons of around twenty *Sinornithomimus*, an ornithomimid whose adults reached 2.5 meters (over 8.2 feet) in length and weighed around 100 kilograms (220 pounds).[13] But the Suhongtu skeletons are nowhere near this size. Most were just one or two years old, with femurs measuring around 20 centimeters (not quite 8 inches). Two individuals must have been three years old and two others seven to eight years old with femurs of about 40 centimeters (15.75 inches) in length. Paleontologists who have worked on this deposit think these animals formed a herd that became trapped in the mud of a pond or lake that was evaporating. The researchers believe the absence of adults and newborns reflects the structure of this population. The young and subadults roamed around together; it's clear that it wasn't safe to leave children unattended. But compared to the young, there were fewer subadults in the group, as in *Psittacosaurus* for that matter, which reveals another characteristic of dinosaur populations: the low number of subadults at any given time. In other words, given the high juvenile mortality rate, there were few seven-year-olds left compared to the number of one- to two-year-olds.

Footprints preserved on a cliff at Cabo Espichel in Portugal tell us that 150 million years ago small sauropods moved around together. There are seven parallel trackways left by animals whose feet measured about 40 centimeters (15.75 inches) long, meaning these animals must have been 7 to 8 meters (23 to 26.25 feet) long. It's perhaps these footprints, or others on the same cliff that's full of them, that inspired the legend of the apparition of the Virgin Mary on a mule, which left no trace but its hoofprints. In 1410, a man, who may have enjoyed some local Douro wine, witnessed Mary emerge from the sea on her mule at Cabo Espichel. Legend has it that the hoofprints of the Virgin Mary's mule are imprinted in the rock. And, as fate would have it, this rock is full of sauropod footprints.

Some sauropods must have also lived in age classes. Given the size difference between adults and juveniles, cohabitation would have involved a great deal of coordination to avoid accidentally crushing their offspring. Since a sauropod egg generally holds 2 to 3 liters (a little more than 2 to 3 quarts), a newborn didn't weigh more than 3 kilograms (6.6 pounds) for a length of around 60 centimeters (about 23.5 inches), whereas its parents weighed several dozen metric tons for a length of 20 to 30 meters (65.6 to 98.4 feet). And since sauropods weren't the most intellectually equipped of the dinosaurs, it was probably best to spare the trouble of educating the young. But for a moment, imagine having to look after a child ten thousand times lighter than its parent, weighing 5 or 6 grams (0.18 or 0.21 ounces) at birth. The Portuguese tracks are far from the only ones to suggest age segregation in sauropods. Other sites, however, such as the ones in Texas, show a mixture of adult and juvenile prints but nothing from newborns. This contradictory data may suggest different strategies existed in different sauropod families.

Recognition Signals

When living in a community, it can be handy to recognize at a glance—or with a sniff—the different individuals who make up the team. For this, the presence of individual recognition signals is useful, whether they're visual, olfactory, or chemical. Nodosaurids' parascapular spines could thus have served to help individuals recognize different members of the same species. As their name implies, these spines were large, somewhat conical osteoderms set on the shoulders that grew horizontally. Parascapular spines have been compared to the horns of bovids,* which biologists consider to be not only a means of defense but also a way for fellow creatures to identity each other. Something like "Right horn 16 inches, left horn 15 inches, 85-degree angle . . . Bingo, it's Margaret!" In *Borealopelta*, the bony core of the parascapular spines is 40 centimeters (15.75 inches) long, reaching 55 centimeters (21.7 inches) with its keratin sheath—far longer than in most modern-day cows. But the horned Barrosã cattle breed from Portugal is even more impressive with a distance between its horn tips of 2 meters (6.6 feet). Among stegosaurs, only *Kentrosaurus* was also equipped with parascapular spines. As it's also the only stegosaur for which sociality has been hypothesized, it all makes sense—*Kentrosaurus*'s parascapular spines could have had the same role of recognition between individuals as that believed for nodosaurids.

Marginocephalians such as *Psittacosaurus* from the Lower Cretaceous of Asia were also gregarious animals. Although this dinosaur is known from a significant number of skeletons, only one of

---

* The Bovidae family includes cows, goats, sheep, and antelopes.

them has "hair" on its backside, preserved with its skin and some strange filaments on the top of its tail. These ninety-one "hairs" extend over approximately 20 centimeters (almost 8 inches) between the second and fifteenth caudal vertebrae. They're grouped in small tufts of three to six individual structures rooted deeply into the skin. The fossil was truncated during extraction, and most of the structures are incomplete; the longest must have been around 15 centimeters (almost 6 inches) long, with a diameter ranging from 0.5 to 1.7 millimeters (0.02 to 0.07 inches). For comparison, human hair is between 0.04 and 0.1 millimeters (0.002 to 0.004 inches) in diameter. Given their thickness, these *Psittacosaurus* structures could be described as "bristles," which is what a pig's thick, stiff hairs are called. *Psittacosaurus* structures also bear a striking resemblance to the spiny structure on the head of the horned screamer (*Anhima cornuta*), a large, noisy South American bird, and to the beard filaments of a male wild turkey (*Meleagris gallopavo*), which has a long tuft of coarse hair that hangs from its breast. *Psittacosaurus*'s bristle-like structures could have had similar characteristics to the structures found in modern birds, but they would have been calcified and highly keratinized, like the horned screamer's spiny structure. But what was the purpose of that little feather duster on its tail? Since *Psittacosaurus* were gregarious animals, perhaps this helped them to recognize each other as individuals, like the manes of certain mammals do. This ridge would have also modified their silhouette, which would then have made them appear more imposing and therefore more intimidating to a predator.

Other obvious signaling structures are present in all marginocephalians, from the domes of pachycephalosaurs to the frills and horns of ceratopsians. And other ornithischians had filaments quite similar to those of *Psittacosaurus*, such as *Tianyulong*, a small heterodontosaurid from the Upper Jurassic of China, which had a ridge of structures from the back of its skull to the beginning of its tail.

Hunting

Hunting, we should remember, is a practice that enables a predator to kill its prey. In the absence of suitable tools (or weapons, if you prefer), the predator must come in contact with its prey using the parts of its own body likely to cause lethal damage, such as its teeth or claws in the case of carnivorous dinosaurs. This requires physical contact between predator and prey, meaning either a chase or an ambush. Proof of this type of behavior is hard to find, since the teeth marks found on dinosaur bones are more likely to be traces of feeding and dismemberment, rather than predation. These marks prove that carnivorous dinosaurs did eat the meat of dead animals, not necessarily that they hunted actively. It isn't only pure scavengers that stumble across the corpse of an animal that has died accidentally and feed on it; just about all predators engage in this practice. In short, a good hunter never turns down a free meal.

Fortunately, as with all predatory species, among dinosaurs there were also unskilled hunters, clumsy ones, ones who missed their target. An American paleontologist discovered a hadrosaur tail vertebra deformed by a large bony callus, a sign of a serious infection. And in the middle of this callus was the tip of a *Tyrannosaurus* tooth. It seems a *T. rex* had attacked this hadrosaur by biting its tail, the herbivore escaped, and then its tail took a long time to heal. This is definitive proof that, while *Tyrannosaurus* may have been a scavenger when it had nothing better to do, it was also an active and sometimes clumsy predator.[14] And since we're on the subject of *T. rex*, while it was happy to eat hadrosaurs and ceratopsians, it also fed on other tyrannosaurs. Traces of large carnivorous dinosaur teeth on *T. rex* bones are evidence of the fact that, as the only large carnivorous dinosaur in its ecosystem, it was a cannibal.[15] This does not prove, however, that *T. rex* killed its fellow tyrannosaurids in order to eat them. It might have just nibbled on their corpses. In present-day large terrestrial predators (lions, Komodo dragons, crocodiles, and so forth), cannibalism is usually the result of predation, so it can't be ruled out that *Tyrannosaurus* slaughtered its fellow tyrannosaurids to feed on their flesh. On the other side of the world, on the large island of Madagascar, another large theropod, the abelisaurid *Majungasaurus*, was involved in the same scenario. Paleontologists working on sites in the Upper Cretaceous Maevarano Formation have found numerous bones bearing the marks of *Majungasaurus* teeth, identifiable by the spacing between the teeth and the small striations made by the teeth's crenulations.[16] These bones belonged to the titanosaur *Rapetosaurus* and to its own species, meaning *Majungasaurus* was therefore another cannibal.

Some theropod dinosaurs hunted, and some were cannibals. As everyone who's seen *Jurassic Park* knows (i.e., perhaps the vast majority of earthlings), raptors hunted in pairs or groups, opened doors, and tapped their claws on stainless-steel kitchen tables. Since raptors were endowed with the largest brains of all the dinosaurs, we're right to expect relatively interesting behavior from them. But are there scientific arguments for believing raptors used hunting strategies?

We already have proof that several species of small theropods lived in packs. To cite just one example with which I'm familiar, sandstone quarries in a geological formation dating from the end of the Early Cretaceous Period can be found on the banks of the Mekong River in northeast Thailand. In the early 21st century near the village of Tha Uthen, numerous footprints of carnivorous dinosaurs and crocodilians were unearthed in one of these quarries.[17] You can follow the trackways of dozens of small predators moving together in the same direction for several meters. Clearly, these animals weren't getting together to play cards. Given their predatory lifestyle, they had to stalk and kill prey. Did they hunt together by encircling their future victim or driving a herd of herbivores toward a few of their congeners? The footprints don't tell us anything about this. For some researchers, the fact that carnivorous dinosaurs lived together does not imply they hunted together. Pointing to present-day Komodo dragons as an example, they suggest that theropods might have converged on a slain animal to compete for the feast, but in no way were they capable of coordinating their efforts to slaughter the animal together.[18]

Small theropods are not the only predators for which we have evidence of sociality. The remains of forty-six large theropod *Allosaurus* individuals were excavated in the 1960s from the Cleveland-Lloyd Dinosaur Quarry in Utah. *Allosaurus* lived during the Late Jurassic Period in what is now the United States. The animals discovered in Utah ranged in length from 3 to 12 meters (9.8 to 39.4 feet), representing a good part of this dinosaur's growth stages. But the way this remarkable deposit was formed—by the accumulation of corpses over the years during a river's flooding—doesn't allow us to conclude that all these *Allosaurus* lived together.

As usual, it's the footprints that tell the tale, even for large predators like the *Acrocanthosaurus* at Glen Rose, Texas. Since a demonstration is never complete until we talk about tyrannosaurs, what about these big beasts? Three short tyrannosaurid trackways were found in British Columbia—short because two of these trackways only show two successive footprints, which is the minimum required to qualify as a trackway, and the third trackway only shows three prints. Seven footprints isn't much, but their arrangement shows unequivocally that they were made by three animals moving in the same direction. These tridactyl-shaped prints reaching up to 67 centimeters (just over 2 feet) in length were produced by large dinosaurs, probably the tyrannosaurid *Gorgosaurus*. So even the brutish tyrannosaurids were social creatures that went about with their friends, no doubt up to no good.

Absence of Proof and Proof of Absence

Finally, while footprints tell us that some theropod dinosaurs were social animals, this notion tells us nothing definitive about their hunting strategies. Since it's always tempting for scientists to shoot down a model promulgated by movies, demonstrating that raptors didn't hunt in packs has become a reasonable exercise for paleontologists. The latest to tackle the issue are American researchers who used carbon isotopes contained in the enamel of fossilized *Deinonychus* teeth to demonstrate that adults and juveniles had different diets.[19] More precisely, the researchers showed that large teeth contain less carbon-13, and small teeth contain more. Adult *Deinonychus* must have fed on the herbivore *Tenontosaurus* (whose isotopic signal is close to that of adult *Deinonychus*), confirming a long-established predator-prey relationship. Juvenile *Deinonychus*, on the other hand, fed on smaller but trophically higher animals. The researchers interpreted this data as evidence of an absence of sociality, with the young not feeding like their parents, bringing the raptors down to the intellectual level of the Komodo dragon. While it seems to be understood that juveniles didn't take part in the parental meal, this conclusion tells us nothing about the hunting techniques used. While the method the researchers devised to answer the question of whether *Deinonychus* hunted in packs was undeniably sophisticated, unfortunately they failed to truly answer it. As their conclusion is, after all, mendacious, I'll summarize it for you: as the isotopic data from *Deinonychus* teeth resemble those of modern asocial reptiles, *Deinonychus* must have also been asocial and therefore not intelligent enough to hunt in packs. But this discovery can be interpreted quite differently in light of what we know about dinosaur sociality: the possible absence of multigenerational groups in *Deinonychus* echoes many other dinosaur clans whose juveniles led life apart in their own youth groups. So, even with all this technology, we're still at the same point: we don't know whether theropods, as social animals, coordinated their hunting strategies to slaughter their prey.

The only dinosaur fossil to show evidence of predation was discovered in Mongolia's Gobi Desert by a Polish expedition in August 1971.[20] In this region, many dinosaurs were fossilized during sandstorms or dune collapses, as the area was also a semidesert at the end of the Cretaceous Period. We therefore find animals in the position in which they abruptly died, often standing or incubating their eggs in the case of certain oviraptorids. Interestingly, a *Velociraptor* skeleton and a *Protoceratops* skeleton were found entangled, seemingly testifying to the attack the raptor had made on the herbivorous dinosaur. The *Velociraptor* had one hind leg wedged under the *Protoceratops* and the other leg apparently stuck in the throat of the herbivore, who was also holding the predator's arm in its beak. This seems to have been a fight to the death between these two creatures,[21] ending in the death of both and followed by their rapid burial in the sand. Because only one *Velociraptor* was involved, this discovery obviously doesn't confirm the hypotheses of pack hunting.

Defense Strategies

Whether theropods hunted alone or in packs, their claws and fangs were grave threats, and the various species of herbivorous dinosaurs tried a variety of techniques to escape them.

The survival strategy of the slow, bare-skinned sauropods was deterrence. Since they weighed more than 10 to 20 metric tons, any aggressors with even a modicum of intelligence knew to avoid a direct attack. But tales of carnivores ravaging the tail of a sauropod before the information had even reached its brain 30 meters (over 98 feet) away are missing a few facts. First of all, sauropods had a well-developed sense of smell and large orbits that suggest good vision; it would have been difficult for a predator to reach the tip of *Diplodocus*'s tail without it being alerted by the smell or seeing the threat. Not to mention the fact that these dinosaurs likely lived in herds. With around thirty eyes and as many olfactory lobes, the herd wouldn't have missed *Allosaurus*'s movements. Last but not least, the legend is based on the speed of nerve impulses in axons (nerve fibers) devoid of myelin, equivalent to 1 to 10 meters (3.3 to 32.8 feet) per second. At 1 meter (3.3 feet) per second, it would take 30 seconds for the nerve impulse to travel 30 meters (98.4 feet), which is plenty of time for voracious carnivores to peck off a few meters of tail before getting a reaction, with the reaction itself delayed by another 30 seconds. But some vertebrate axons are surrounded by a myelin sheath that isolates them, making the speed of signal propagation a different order of magnitude: around 50 meters (164 feet) per second. This leaves *Allosaurus* with time for only half a mouthful before taking a monumental slap in the face—clearly not a good strategy.

Large sauropods probably had no predators once they reached a critical size. On the other hand, as they would eventually die of natural causes or accidents, their corpses became an enormous source of protein for countless hungry animals, from the largest carnivores to the necrophagous insects that finished the job. In a controversial but rather amusing study, ecological researchers theorized that the number of sauropods dying every day at the end of the Jurassic Period in the United States would have been enough to feed the entire *Allosaurus* population.[22] And so *Allosaurus* could have been content to be a pure scavenger like vultures today. This theory is based on the extraordinary biodiversity of large sauropods in the Morrison Formation, a geological formation deposited between 156 and 147 million years ago in what is now the western United States. Nearly twenty sauropod species, including *Diplodocus*,

Camarasaurus, and *Brachiosaurus*, have been described in this rock, and several of these were found in the same deposits and were therefore clearly contemporaries. Assuming a density of one to two giant sauropods per square kilometer (0.4 square miles), the ecologists' models provided enough tons of dead meat to feed entire populations of allosaurs. This is of course theoretical, and other data shows that allosaurs must also have been active predators, but this work has the great merit of raising questions about the ecosystems fed by these sustainable food sources (the bodies of very large animals). The corpses of blue whales that sink to the bottom of the ocean feed hundreds of species for decades. While the decomposition process is quicker in the open air, it probably still took months or years for land animals to completely graze on a gigantic sauropod corpse. And the whole guild of predators, from the largest to the smallest, had something to gain from it.

Psittacosaurus are among the dinosaurs that aren't particularly fast, heavy, or covered in protective armor. Their skeletons, which measure 1 to 2 meters (3.3 to 6.6 feet) long, are often found in large numbers, so they probably lived in large herds. A gregarious lifestyle is already a first line of protection against predators who are likely to choose an older or sicker animal. To confuse the attacker, *Psittacosaurus* had another strategy recently revealed by the study of a specimen with preserved skin.[23] This meticulous study has shown that the shade of the preserved skin varies according to area of the body. The ventral part (under the stomach and tail) of the animal was pale (with the exception of the cloacal vent, discussed in chapter 5), while the dorsal part was more strongly pigmented. The presence of melanosomes shows that the dominant color was brown to reddish-brown. Having a lighter underside and darker upper side is a common characteristic of modern animals and is a camouflage strategy known as Thayer countershading. "The principle behind this camouflage method is a gradual gradation of the animal's coloration, with the parts that are usually the brightest also being the most colorful, and vice versa. Because they contrast less, areas of light and shadow tend to blend into one another, making the animal less visible."[24] In open habitats (such as savannas), the light-colored part extends farther up the animal's sides, whereas in forest habitats, only the ventral part is depigmented. This was true of *Psittacosaurus*, which was quite dark on the sides, so it's likely that it frequented enclosed habitats and didn't get much sun.

Exactly the same strategy, albeit with a few color nuances, has been found in the nodosaurid *Borealopelta* from the Lower Cretaceous of Alberta, Canada.[25] This remarkable fossil, covered with plates and bony spines, must have been as edible as a tin can. In particular, it had long parascapular spines on its shoulders, which must not have made it too appetizing to its carnivorous cousins. In addition to this carapace, *Borealopelta*'s color distribution was similar to that of *Psittacosaurus*, with an underside lighter than its back—enough to further annoy visual predators. A present-day mammal of equivalent mass is the rhinoceros. It has dispensed with gradation of colors and is instead uniformly gray since it's theoretically devoid of predators, with the exception in recent decades of a few deluded primates convinced that finely ground rhinoceros horns will restore their failing virility. But in *Borealopelta*'s day, weighing a ton and being caparisoned in bone may not have been enough to intimidate theropods in search of a snack. *Borealopelta*'s large parascapular spines, as their name suggests, extend from their shoulders and are oriented laterally. These spines are found in several nodosaurid species, such as *Polacanthus*, and as discussed earlier, the function of these structures is now being compared to that of bovid horns, which are used for recognition between individuals.

Ankylosaurids' defense strategy was similar to their nodosaurid cousins'. Their skin was entirely encrusted with osteoderms to break predators' teeth, and they had a bony club at the end of their tail. The club was formed by osteoderms fused to the tail's last vertebrae. It could weigh several dozen kilograms in the largest species, and wasn't just a perch for small birds and tired pterosaurs. Paleontologist Victoria Arbour has calculated the damage that could have been caused by the infamous club. A large ankylosaur like *Euoplocephalus*, with a tail measuring 3.5 meters (almost 11.5 feet) long with a mass weighing 20 kilograms (44 pounds) at its tip, would have had an impact force of 36,000 to 72,000 N/cm^2 (newtons per square centimeter) and an impact stress of 364 to 718 MPa (megapascals)—more than enough to pulverize a bone![26] So the club wasn't just for show. You'd have had to be really hungry to consider making a meal of a *Euoplocephalus* or *Ankylosaurus*.

The ceratopsian family is easily recognized by its disproportionately large skulls with spiky frills. There is speculation they may have used them to impress their tyrannosaurid predators, but their effectiveness may have been rather limited, since the only known *Tyrannosaurus* coprolite is stuffed with small pieces of bone from juvenile ceratopsians[27] (and hadrosaurs). Adult *Triceratops*, reaching 9 meters (29.5 feet) in length and weighing around 10 metric tons, must have been less worried. As for the famous *Triceratops* horns, we've all seen depictions of them making *Tyrannosaurus* retreat.

In any case, the presence of lesions on *Triceratops* frills suggests the possibility of inter-congener fighting.[28] In other ceratopsians, such as *Centrosaurus*, such lesions are rare, which suggests that frills and horns were more for show or, if you prefer, must have been useful for visual communication within the species. In summary, most ceratopsians were rather large, rather gregarious, and had scary-looking heads—a set of characteristics that probably enabled them to live well in their forced cohabitation with tyrannosaurids.

Many hadrosaurs also had to cohabit with these nasty beasts, and they were clearly less well armed than the ceratopsians. Not being big enough to defend themselves, not having bony armor or horns, what did they have to use against a tyrannosaur? According to some paleontologists, they loved to travel! You might even say they were globetrotters. Known as the "caribou of the Cretaceous," this hypothesis had them traveling thousands of miles every year between the far north of the American continent and more temperate latitudes. Until recently, this proposition explained the presence of dinosaur fossils in very northerly regions such as Alaska and northern Canada. Although the climate was warmer than it is today, it was still dark for six months of the year in the Arctic Circle. Hadrosaurs of the genus *Edmontosaurus*, the hypothesis goes, would simply have spent the summer in these hostile places before returning to milder climates. Like so many species today, some dinosaurs would have been migratory. But this appealing hypothesis was shattered following further investigations into the geochemistry of hadrosaur teeth. If a dinosaur travels several thousand kilometers or miles every six months, we should be able to find the isotopic signature of these journeys in its teeth. Analysis of strontium isotopes contained in the teeth of North American hadrosaurs has invalidated this idea of long-distance migratory behavior.[29] Hadrosaurs moved—but not great distances. One of the individuals studied moved between Alberta and Saskatchewan, Canada, a distance in the order of 80 kilometers (about 50 miles). This is comparable to the short migrations of elephants but not at all to those of caribous, which cover more than a thousand kilometers (over six hundred miles) from north to south and south to north every year.

So if Alaska's *Edmontosaurus* didn't migrate south as winter approached, how did they endure the polar night? While we don't know the answer to this question, we do know that they remained there for the winter. This is demonstrated not only by their bone tissue, which alternates between periods of weak growth (during the polar night) and strong growth (during the summer), but it's also demonstrated by the discovery of a significant number of young individuals who wouldn't have been able to take part in long seasonal migrations. Hadrosaurs therefore lived in the Arctic Circle, where they certainly experienced snow and ice, despite it being a generally warmer climate than it is today. It's hard to imagine bare-skinned animals surviving the blizzards of the far north, and since they didn't make igloos, we're right to wonder. Were they covered in hair, like the first mammoth? Unfortunately, we have yet to find the skin of a woolly hadrosaur.

One element seems to bring *Edmontosaurus* closer to its distant great-uncle *Dysalotosaurus*: a recent study of this polar hadrosaur's footprints revealed they belonged to a mixture of individuals ranging in age from the very young, with footprints measuring 10 to 20 centimeters (almost 4 to 8 inches) long, to adults with footprints approaching 70 centimeters (27.5 inches). The discovery of thousands of footprints in Alaska's Denali National Park strongly suggests the existence of large herds that included all age groups.[30] It also shows that a polar world, in times of greenhouse effect, can be home to vast animal populations—information that could well be useful in the near future.

We still haven't answered the original question, however. How did *Edmontosaurus* cohabit with the gluttonous *Tyrannosaurus*? Juvenile hadrosaurs were part of the monster's daily calorie intake. According to researchers in Alberta, Canada, who reconstructed *Edmontosaurus*'s musculature, perhaps there was hope for them after all.[31] Adult hadrosaurs would have had more endurance than *Tyrannosaurus*, who was more of a sprinter, built for speed. Keeping your 8-metric-ton body moving really fast over a long distance is not recommended by cardiologists, especially when your diet is totally unbalanced. Tyrannosaurids weren't the only theropods with good acceleration. In a quarry in Oxfordshire, British paleontologists discovered the trackway of a theropod that was 2 meters (6.6 feet) tall at the hip[32] and must therefore have been 5 to 6 meters (16.4 to almost 20 feet) long. The prints of two of the animal's successive right feet started close together (only 2.7 meters [almost 9 feet] apart) and then spread farther apart, until two successive right feet were almost 6 meters (almost 20 feet) apart. Translation: this theropod accelerated, and not by just a little. In a few dozen meters, it went from 6.8 km/hour (4.2 mph) to 29.2 km/hour (18.1 mph)—from a slow trot to full gallop! This is roughly the average speed of the world-record holder in the 800-meter race (28.51 km/hour, or 17.7 mph) and not much slower than the speed of the best 400-meter runners (32 km/hour, or 19.9 mph). In any case, it's far too fast for most of us and for most dinosaur species to keep up! These theropods in Oxfordshire were from the Bathonian Age, around 165 million years ago. Much later in Texas, 110 million years ago, two other medium-sized theropods were speeding around even faster,[33] between 40 and 43 km/hour (about 25 to 27 mph), which is much faster than Usain Bolt himself (37.58 km/hour [23.4 mph] over 100 meters, peaking at 44 km/hour [27.3 mph] between 60 and 80 meters). Other theropods found in the United States in 2021 broke the record with speeds exceeding 50 km/hour (31 mph). These were small animals from the Early Jurassic Period, nearly 200 million years ago, moving at 54 km/hour (33.6 mph). One foot measured 33 centimeters (13 inches), making a dinosaur 1.3 to 1.5 meters (4.3 to 4.9 feet) tall at the hip, with a stride (defined here as the distance between two successive right feet) of over 7 meters (23 feet).[34]

While the hadrosaurs relied on long-distance running as a strategy to lose *T. rex*'s interest, other dinosaur species used sheer speed to dissuade their predators. Unfortunately, these accelerating footprints don't tell the whole story since the vast majority of the millions of known footprints are from dinosaurs moving at a walking pace. The Oxfordshire megalosaur and its buddies in Texas and Utah are the exceptions that prove the rule. This suggests that things weren't moving any faster in the Mesozoic Era than they are today. For example, if you observe animals in the African savanna, or just about anywhere else far from concrete, it's clear that for the most part herbivores graze and carnivores sleep. It's true, observing nature isn't always exciting. If we calculate the speed of the dinosaurs whose footprints we know, the results aren't too impressive— between 0.5 and 10 km/hour (0.3 to 6.2 mph), not even fast enough to escape an old *Megalosaurus* with one working eye and a limp! In the absence of running footprints of these dinosaurs, we fall back on models to give us an idea of their maximum speed, based on the proportions of their limb bones. And when it comes to paleontology, models can often mean anything. In terms of maximum speed, published estimates vary from 5 to 20 meters per second for a tyrannosaur, which equates to 18 to 72 km/hour (11.2 to 44.7 mph). Such scattered results are a sign that something is amiss, and so these models are of little use. Staying with anatomy, we can see that animals like ornithomimids, whose femurs are shorter than their tibias (like ostriches), must have been excellent runners, but trying to quantify their performance is beyond our means. By the way, if you're wondering how zoologists know that an ostrich can reach 60 or 70 km/hour (37 to 43 mph), well, it's very simple. The great zoologist Robert McNeill Alexander, for example, drove a big 4×4 next to them for two or three minutes while watching the speedometer. Naturalists filmed the ostriches at the same time to try and fine-tune the speed afterward.[35] Although we'd like paleontologists to give coherent estimates of dinosaurs' maximum speeds, unfortunately we don't have galloping dinosaurs to observe and big cars speeding right along next to them!

Parasaurolophus's Dietary Supplements

What we know today about dinosaur brains doesn't tell us much about their dietary preferences. So to study this topic, we rely on the classic formula from comparative anatomy: sharp teeth + claws = carnivore. And the rest would have been herbivores (or vegetarians, as grasses and therefore "herbs" didn't appear until the Cretaceous Period) or even omnivores, but we don't really know for sure. In any case, it's certain that feeding must have been a fairly constant preoccupation for dinosaurs, given their physiology. Being warm-blooded animals, the carnivores had to eat almost daily, and vegetarians continuously. While a Komodo dragon can wait a month between meals, this was not the case for a tyrannosaur. The two most valuable sources of information on dinosaur diet are the rare stomach contents preserved in situ and the equally rare coprolites (or fossilized feces). A huge North American coprolite, thought to be the product of *Tyrannosaurus rex*, was full of bone fragments of juvenile hadrosaurs and ceratopsians. This gluttonous *T. rex* crushed bones in its powerful jaws and swallowed everything. And it had a particular fondness for juvenile hadrosaurs and ceratopsians.

A recent discovery by American paleontologist Karen Chin concerns the diet of ornithopods (the group containing hadrosaurs and *Iguanodon*), which were presumed to have been strictly herbivorous,[36] although some of last century's fiction may have suggested otherwise. In 1936, for example, an *Iguanodon* savagely attacked a young girl on a lost island in a comic strip by artist and scriptwriter Nicolas Mengden, who was also a Russian count in exile.[37] But beyond these fantasies, scientists have always considered hadrosaurs to be strict herbivores—and perhaps quite wrongly so! If you think about it, today's herbivores rarely have as strict a diet as a vegan human; giraffes will happily nibble on bones, while hippopotamuses or deer may be tempted by a piece of meat. Not to mention the involuntary ingestion of pebbles when swallowing a bush with its roots. A little phosphorus or calcium is always welcome. Food supplementation is therefore common in nature today, and we'll see that it was just as common seventy-five million years ago. The coprolites studied by Karen Chin were discovered in Utah. Some of them were quite large and grouped together in what paleontologists—like many others—call "latrines," places where dinosaurs regularly went to poop. Evidence of paleolatrine behavior is extremely rare to discover, but as in archaeology, coprolites come with their share of surprises. One of the main components of the coprolites found in Utah was fossilized wood. The defecating dinosaur would have swallowed large mouthfuls of decaying wood—meaning, wood degraded by fungus, which made it more edible and cellulose rich. More astonishingly, these paleofeces were filled with fragments of crustacean cuticle. This hadrosaur (probably *Parasaurolophus* and possibly *Gryposaurus*) supplemented its diet by devouring numerous crustaceans of at least 5 centimeters (about 2 inches) long. These were probably not accidentally ingested while eating a fern or a piece of wood but were prey that had been purposefully selected. Was it because they tasted good? Or because they provided a number of essential nutrients, such as phosphorus or calcium? Maybe both, but one thing's for sure: hadrosaurs happily ate crawfish, or at least their Cretaceous cousins.

Coprolites should not be confused with cololites. Cololites are fossilized stomach or intestinal contents and are even rarer than coprolites. As described in 2020, a cololite from *Borealopelta*, a 125-million-year-old ankylosaur from what is now Canada, contained 85% fern fragments, some cycads, conifers, and charcoal.[38] *Borealopelta*, a member of the family Nodosauridae, seems to have had a selective diet, concentrating on one fern species in particular. And like modern herbivores, it gravitated toward recently burned forests, where ferns are the first to grow back, which would explain the presence of charcoal in its stomach. It's worth noting that even in the absence of bipedal pyromaniacs, forests burn all the more when oxygen levels in the atmosphere are high. The air in Earth's atmosphere is currently 21% oxygen, but this has varied between 15% and 33% over geological time. At certain periods, such as the end of the Carboniferous Period (when oxygen levels were over 30%), even damp vegetation could catch fire like tinder. In contrast, below 19% oxygen makes the probability of fires very low.[39] Much to *Borealopelta*'s delight, it seems that oxygen levels were higher in the Cretaceous Period than they are today, leading to numerous forest fires. Back around 100 million years ago, there were even peaks of 30% oxygen—a permanent fireworks display!

Tongue-Tied

Still on the subject of ankylosaurs, investigations by American paleontologists into the skull of a *Pinacosaurus* from Mongolia could have some rather fascinating outcomes. The researchers were able to identify a well-ossified hyobranchial apparatus in this dinosaur.[40] In almost all dinosaurs known to have had hyoid bones, they were simple in shape and included just two bones, like in crocodilians; so, we deduce from this that their tongue must have been fairly basic and glued to the palate. There's no danger of *T. rex*'s tongue bursting out of its mouth to take you away like in *The Lost World*! On the other hand, *Pinacosaurus*'s hyoid apparatus comprises at least eight ossified elements, suggesting the presence of a powerful, muscular tongue—the same type of adaptation found in many insectivorous animals, from salamanders and pangolins to chameleons and anteaters. The presence of numerous tongue muscles attached to the hyoid bones may well be evidence of entomophagy (insect eating) in *Pinacosaurus*, which would have devastated Cretaceous termite and ant mounds. It's a tempting hypothesis, but unfortunately we'll have to wait a little longer to be certain, as only the discovery of a *Pinacosaurus* cololite full of insect cuticles will be able to validate it definitively.

Finally, even more rare than cololites and coprolites, and just as useful for learning about a dinosaur's diet, are regurgitalites, which are, as the name suggests, fossilized vomit remains. The discovery of small piles of tiny mammal bones in Montana composed of indigestible parts, such as teeth showing traces of corrosion due to digestive juices, suggests that some dinosaurs regurgitated gastric pellets like present-day birds of prey.[41] In this case, the predator was *Troodon*, a small carnivorous dinosaur, and the victims were marsupials named *Alphadon*. So *Troodon* ate *Alphadon* and spat out the most indigestible bits, just like the first owl.

5
Banter Between Lovers

Markers of cognitive ability in the animal world include all aspects of reproduction, from courtship rituals and nest surveillance to the care of young. The discovery of a large number of fossils over the last thirty years has led to the development of many hypotheses on all of these subjects. One of the main questions concerns the recognition of males and females—or in other words, whether there was sexual dimorphism in dinosaurs. And as we shall see, this isn't easy at all to determine. It's difficult enough to recognize the difference between males and females in modern crocodilians, with females generally being only a bit smaller. An exception is the gharial, or fish-eating crocodile, since adult males can be easily identified by the presence of a large protrusion on the snout, but in the case of the other twenty-four crocodilian species, it's much more complicated. As for dinosaurs, the first obstacle in recognizing a morphological variation due to sex is the scarcity of material; as most species are known only from a few bones or a single skeleton, it's unrealistic to expect to find any variation in them. For the few species known from numerous skeletons, such as *Protoceratops*, *Tyrannosaurus*, *Coelophysis*, and *Kentrosaurus*, several studies in the late 20th and early 21st centuries suggested fairly understated sexual dimorphism. But then wham! A recent statistical review of these studies has invalidated them all. No one has statistically demonstrated the existence of sexual dimorphism in any dinosaur species.[1] This is not to say that there were no differences between males and females, but simply that the samples we have available don't allow us to draw any conclusions. No doubt to avoid disheartening his readers, the author of this study explained the only way to truly launch a sexual dimorphism analysis is to start with specimens whose sex is known for sure, such as fossilized animals with eggs inside their skeleton (a known case: a *Sinosauropteryx*, a small carnivorous dinosaur from China) or with medullary bone, a type of bone tissue known only in female birds at the time of egg laying (a known case: a *Tyrannosaurus*). In short, there's still a lot of work to do. For the moment, there are no known examples of sexual dimorphism in dinosaurs, but that doesn't mean we can't take a closer look at their sexual relationships.

Courtship Rituals

Osteological clues to intraspecific interactions (the term used to describe two pachycephalosaurs, for example, headbutting each other) abound among dinosaurs, and paleoichnologists have put forward some interesting hypotheses, which we're going to describe in detail next. From confrontations between males to courtship rituals just prior to mating, modern birds display a diversity of extremely complex behaviors, some perhaps inherited from their distant ancestors.

HEADBUTTING BEHAVIOR

When an anatomical feature as bizarre as the presence of a 25-centimeter (9.8-inch) compact bone mass on the top of the skull is favored by natural selection, you'd think it must serve some essential purpose. The trick, then, is to understand whether pachycephalosaurs, in this case, were living at the foot of cliffs from which pebbles regularly fell, providing an explanation for their helmets. Admittedly, this isn't the preferred explanation for this strange structure. Pachycephalosaurs, with their distinctive bony skull dome, have long been suspected of having been the rams of the Cretaceous, using this bony mass to settle disputes with headbutting.

In 1955, the American paleontologist Edwin Colbert was the first to suggest this possibility, which was so novelistic that the following year it appeared in a short story by the American science-fiction writer Lyon Sprague de Camp. The time-travel story "A Gun for Dinosaur" takes a close look at the equipment needed to hunt dinosaurs in the Cretaceous Period thanks to a time machine. The narrator (a hunting guide) advises his clients never to aim for the brain when shooting prey: "People used to hunting mammals sometimes try to shoot a dinosaur in the brain. That's the silliest thing you can do, because dinosaur haven't got any. To be exact, they have a little lump of tissue about the size of a tennis ball on the front end of their spines, and how are you going to hit that when it's imbedded in a moving six-foot skull?" Like any self-respecting Cretaceous safari, the story ends with a fatal encounter with a tyrannosaur on the prowl, but in the meantime the hunters talk about the habits of pachycephalosaurs before killing one: "These are the troodonts, small ornithopods about the size of men with a bulge on top of their heads that makes them look quite intelligent. Means nothing, because the bulge is solid bone and the brain is as small as in other dinosaur, hence the name. The males butt each other with these heads in fighting over the females."[2]

Debated initially, this hypothesis has been revived following new studies that show the presence of cranial-bone lesions fully compatible with this unfriendly attitude.[3] Flat-skulled pachycephalosaurs (lacking the large bony dome) are now considered to be the females or more likely the juveniles, neither of whom would have been expected to engage in combat behavior. As for the way in which pachycephalosaurs settled scores, an American paleontologist has pointed out that the geometry of the skull was not ideal for a head-to-head confrontation. Instead, he considered the possibility of head blows into the opponent's flanks, which must have caused lasting pain. More recent studies, however, suggest the possibility of violent head-to-head matchups. Given these recent findings, these magnificent bony domes were probably not just for show but to settle disputes between males.

Other species also show lesions that may have been due to violent contact with members of the same species, such as certain ceratopsians, particularly *Triceratops*. Numerous lesions have been found on *Triceratops*'s frill and seem likely to have been caused by blows from others' horns. *Triceratops* had two large supraorbital horns on its brow and a small horn on its snout. Its cousin *Centrosaurus* had the opposite: a large horn on the snout and two small horns above the eyes. *Centrosaurus* has shown virtually no frill lesions, perhaps a sign it was as gentle as a lamb, or at least much less quarrelsome than *Triceratops*. But this is far from certain because, although *Centrosaurus* didn't engage in head-to-head fighting, fossils of both it and *Pachyrhinosaurus* (another ceratopsian living in large herds) reveal a large number of broken and healed ribs. So *Centrosaurus* seems to have been more adept at flank-butting—meaning, it used its head to butt into its rival's flanks like the American bison.[4]

In any case, marginocephalians seem to have loved fighting. Traces of intraspecific fighting are frequent in many species of this large Cretaceous group and can be interpreted as either territorial behavior or clashes between males in order to mate with females, or between females to settle disagreements.

KISSES FOR *T. REX*?

Tyrannosaurs, of course, were rougher than other dinosaurs, and even when it came to love, they lived with gore close at hand. It might well have been that these animals, whose breath must have rivaled the jackal's, viciously bit each other's snouts to the point of leaving deep tooth marks in their competitor's jawbones. An in-depth study of hundreds of tyrannosaurid jaws has concluded that around half of all adults show this type of lesion.[5] But when we say "half," we immediately wonder: males or females? The authors of this study see evidence of a male behavior like that of crocodilians: during the breeding season, face-to-face combat would have established a hierarchy of males for access to females. Tyrannosaurs are not the only dinosaurs to show these facial lesions; a good number of other genera of large theropods, such as *Allosaurus* and *Carcharodontosaurus*, also do. Carnivorous dinosaurs therefore seem to have been accustomed to bloody jousts with their congeners. Dromaeosaurs and other birdlike theropods, on the other hand, have no such scars, which might suggest these feathered animals had moved on to other tactics, perhaps with courtship rituals more like those of birds than the bloody biting of their cousins.

THE THEROPOD LEK

Some modern bird species, such as plovers, are fond of courtship displays during the mating season in which males scrape the ground, mimicking the digging of a nest, thus increasing their chances of being chosen by a female. In the New Zealand parrot, ornithologists have named these areas in which males gather for competitive courtship rituals a "lek." Paleoichnologist Martin Lockley has described similar scrape marks in Colorado, but these ones are fossilized and on a completely different scale. These traces date back to the end of the Early Cretaceous or beginning of the Late Cretaceous Period, one hundred million years ago. They resemble vast depressions and measure up to 2 meters (about 6.5 feet) in diameter, with deep claw marks caused by powerful tridactyl legs—in short, the legs of large carnivorous dinosaurs. A closer look at these strange fossils reveals the scratches made by two legs positioned side by side. The animal must have been standing on its feet and alternately digging its claws into that spot on the ground: left! right! left! right! This motion of kicking the sand behind itself created a depression up to 20 centimeters (nearly 8 inches) deep. But the creature wasn't digging its grave or a shelter for the night. Was it digging a nest? Again, this is unlikely as there wasn't any trace of an eggshell at those sites.

What hypothesis remains to explain these scrape marks in Colorado? Researchers believe these deep cavities can be attributed to theropodian courtship rituals and, more specifically, to plover-like "scraping ceremonies." So, instead of ferociously biting the jaws of their rivals to discourage them from competing for females, big theropods would have furiously scraped the ground to impress females and improve their chances of mating.

The Age of Consent

These courtship rituals involved adult animals of reproductive age. But at what age did dinosaurs first make love? Is it possible to seriously answer such a question? It's generally accepted that dinosaurs experienced a decline in growth rate as they reached sexual maturity. From then on, most of their energy would have been devoted to reproduction and no longer to growth. We just need a method to determine this growth rate. Paleontologists have been aware for some time that many dinosaurs have lamellar bone. When cut, lamellar bone shows concentric circles comparable to the growth rings found in trees. These circles correspond to the annual growth arrest lines. All you have to do is spot where the circles are getting closer and closer together (and therefore growth was getting slower and slower), then count the number of circles deposited before this growth arrest, and you've got the dinosaur's age at first flirtation, or first lek. For *Psittacosaurus*, it was eight to ten years old. It was fourteen to fifteen years old for *Apatosaurus*, eighteen years old for *Tyrannosaurus*, and eighteen to nineteen years old for *Camarasaurus*.[6]

LIVE FAST, DIE YOUNG

The age of a dinosaur at the time of its death can be determined by the same method of bone cutting and counting the growth arrest lines. Like James Dean, "*T. rex* lived fast and died young," according to American paleontologist Gregory Erickson, a tyrannosaur specialist. For example, the Field Museum's Sue, one of the largest known *T. rex* specimens, died at the age of twenty-eight, weighing 6 metric tons and measuring 12 meters (over 39 feet) in length.[7] This doesn't mean that a *Tyrannosaurus* wasn't able to live past the age of thirty, just that the largest skeleton specimen discovered to date kicked the bucket at twenty-eight. And I'll remind you that Sue had been very ill before dying. While this dating method provides the age at death of a particular individual, it doesn't provide information on the maximum lifespan of a species. Before bone histology was used to determine the age of dinosaurs, age was ascertained on the basis of growth rates in crocodilians and turtles (which are low), and so giant sauropods were predicted to live for centuries. Unfortunately for sauropods, the data now available isn't so optimistic. According to one study, sauropods were found to have died relatively young,[8] at forty years old for a *Camarasaurus*, between twenty-six and thirty-one for an indeterminate mamenchisaurid, between twenty-nine and thirty-one for an *Apatosaurus*, and so on. All in all, these large dinosaurs were perfect examples of James Dean and Gregory Erickson's motto.

BANTER BETWEEN LOVERS

Mating

There's no doubt that dinosaurs mated once they reached sexual maturity, but a few questions remain about the fundamentals. Did males have a penis? Did females have a clitoris? What position was necessary for penetration? Once again, there are answers to these questions! No dinosaur penis is known to date. Baron Nopcsa's hypothesis in 1905 about the existence of a penile bone in *Diplodocus*[9] was a flop; the bone in question was probably an interclavicle (a pectoral girdle bone). In the absence of a fossil penis, what can crocodilians and birds tell us about this delicate subject? Most male birds, like roosters, have no penis; their sperm is transmitted by cloaca-to-cloaca contact. Other birds, like ducks and ostriches, have a penis tucked away at the bottom of the cloaca, popping out when needed. In any case, penises and testicles are internal organs in both crocodilians and birds. Incidentally, the male duck's penis is shaped like a corkscrew, as is, happily, the female duck's vagina. Crocodiles, however, have a curious claw-shaped penis, and dinosaurs probably did too. The ancestral state of archosaurs therefore includes the presence of a penis in the male, since crocodilians and some birds have one. An important detail: while in placental mammals the urethra carries urine and sperm through the phallus, in crocodilians and birds the urethra opens directly into the cloaca. The sperm is then carried by a spermatic duct from the testicles (also internal) to the penis, flowing above it along a sperm groove until the cavernous body swells during mating and directs the groove into a seminal canal.

As for female dinosaurs, similar questions, similar answers. There is no known fossil clitoris and no known baubellum (clitoral bone of certain mammals), just as there is no known baculum (penile bone) in males. But female crocodiles have a clitoris, which is similar in shape to the penis,[10] and cassowaries and ostriches have one too. So, since the ancestral state of archosaur females includes the presence of a clitoris, female dinosaurs must have had one, just as they had a vagina opening into their cloaca, which their companion's phallus used to deposit sperm.

The Interior of a Dinosaur's Cloaca

Soft tissues are rarely preserved in dinosaurs, and a preserved reproductive organ is certainly a holy grail in paleontology. Researchers recently took a step closer in their search by examining an incomplete dinosaur cloaca for the first time.[11] It belonged to an exceptionally well-preserved *Psittacosaurus* skeleton with skin and strange structures on the top of the tail—a kind of coarse hair forming a small crest, as described in chapter 4. This unusual fossil, preserved in the Senckenberg Natural History Museum in Frankfurt, Germany, after being illegally exported from China, also features melanosomes, which have enabled scientists to reconstruct its color. Better yet, scientists also inspected the partially preserved cloaca. They found the opening of the cloaca is right where we thought it would be behind the posterior end of the ischium, the bone of the pelvis that projects backward under the seventh vertebra of the tail. This opening is surrounded by two lobes that are strongly pigmented with melanin, which could have contained musk glands as in crocodiles. While the Frankfurt *Psittacosaurus* specimen is exceptional, it only shows the cloacal vent and not its internal structure. For more information, we'll have to wait for the Galahad of paleontology to bring us a fully fossilized cloaca.

THE DINOSAURIAN *KAMASUTRA*

All that's left now is for this penis and vagina—that we know nothing about—to meet, with the female's long tail being the main obstacle to this happy encounter. To achieve fertilization, either the two cloacae needed to be brought as close together as possible, or the penis had to be long enough for penetration. Paleontologist Timothy Isles has devoted a long seventy-plus-page article to this fascinating subject, detailing the preferred sexual positions of birds, crocodilians, elephants, and lizards (which, incidentally, have two hemipenises that they use alternately) before tackling the dinosaurian *Kamasutra*.[12] Since birds' tails don't pose an obstacle and crocodilians mate in water, they're not useful models for understanding dinosaurian practices. But Isles refutes the hypothesis of dinosaurs having had a very long penis, because if they had, it would have been supported by numerous muscles like in the elephant, and traces of these would have been found on the pelvic bones. There are no traces of super-penis muscle attachments on dinosaur pelvises though, so it's likely males couldn't penetrate females from a distance. The only other option is that dinosaurs had to bring their cloacae close together. The most believable position would be for the male to pass a hind leg over the female's tail and bring his cloaca close to hers, placing all his weight on her. In support of this scenario, a recent study showed a large number of healed vertical and diagonal fractures on the upper caudal neural spines of hadrosaurs. These fractures are due to a heavy mass pressing vertically, or slightly obliquely, on the damaged tail, most likely from mating.[13] The author tentatively suggests that hadrosaurs with these injuries were females, which seems reasonable. But any conclusions are far more ambiguous when it comes to explaining the presence of fused vertebrae toward the middle of the tail in some sauropod dinosaurs. According to some researchers, this is an adaptation in males that enabled them to put their heavy tails on the ground when mounting females. But others think it's an adaptation in females that enabled them to move their tails aside to facilitate penetration. For large sauropods weighing several tens of metric tons, our dinosaurian sexologist Isles considers the possibility of aquatic mating, while conceding that by the time some sauropods reached maximum size, they may have already passed reproductive age. Finally, we have to talk about stegosaurs. According to Isles, it's clear that their rows of plates and dorsal spines seem to "conspir[e] to render copulation unworkable." He adds, "Considering the distance between both male and female cloaca in a pair of *Stegosaurus* with the male mounted on two legs, a phallic structure would have to exit his cloaca, negotiate around the pubis and then rise upwards to meet and penetrate the female vent. Such a lengthy and curvilinear organ would undoubtedly have required some means of a support configuration, but there is nothing to be gleaned from anatomical evaluations to fit such a purpose." There is also no evidence of a baculum (penile bone). Once the possibility of a super-penis has been ruled out, the remaining hypotheses concerning stegosaurian sexual positions are these: back-to-back coitus, suggested by paleontologist Kenneth Carpenter; or the female lying on her side, a method advocated by the famous Robert Bakker (immortal author of "Dinosaur Renaissance," which revolutionized our understanding of dinosaurs). As you can see, there's never a dull moment with paleontologists. (Well, there are still days when we get a little bored.)

DINOSAUR EROTICA

When scientific certainty is lacking to this extent, it's tempting to consider what novelists and filmmakers might have come up with. Unfortunately, dinosaur coitus is rarely depicted in books or films, no doubt to spare the sensibilities of the young audience fond of these entertainments. So, despite producing six films in a quarter of a century, the directors of the *Jurassic Park* series have yet to capture their protagonists (neither the humans nor the dinosaurs) making love on film. But years before, on June 2, 1922, their predecessors in a different part of the entertainment business were less timid, when at the annual dinner of the Society of American

BANTER BETWEEN LOVERS

Magicians in New York, Sir Arthur Conan Doyle showed a film depicting a scene of dinosaurs making love that titillated the American journalists in attendance.[14] This scene was to be part of the wonderful 1925 film *The Lost World*, based on Conan Doyle's novel of the same name. In the end, it was apparently cut by the censors, and the original footage doesn't seem to have survived.

One of the few scenes of dinosaurs making love in cinema that I know of comes from Picha's slightly madcap cartoon *The Missing Link*, released in 1980. In it, we witness the doggy-style mating of two stegosaurs, culminating in the horrific spectacle of the male being cut in half lengthwise by the female's bony plates. Although we have little faith in the authenticity of this scene, given that stegosaurs survived for several tens of millions of years on Earth, the scene nevertheless forces us to question the practicalities of coitus in this species whose back is studded with vast bony structures that practically forbid the sexual positions common to mammals and reptiles.

Examples of love scenes in literature are also relatively sparse. Instead it's the rather violent relationship between humans and dinosaurs that weaves the narrative fabric of science-fiction novels, with the hunt for dinosaurs (or conversely the hunt for humans) as the principal plot. Dinosaurian sensuality had at least one follower, however: Fernand Mysor, a French novelist and songwriter who is somewhat forgotten today. In his 1923 book, *Les semeurs d'épouvante* (Sowers of terror), the two main characters, Lucienne and Hubert, find themselves back in the Jurassic Period by rather mysterious means, and they are about to meet a tragic fate. On one of their many adventures, they come across two *Atlantosaurus*, each about 30 meters (over 98 feet) long:

> Their skin, streaked with rivulets of gold, was now quivering from head to toe; and apparently to play while awaiting the frightful

embrace, the female lay down on the warm grass. They both roared continuously, entwining their paws and clawing at each other with their lizard-like fingers; and the shuddering that agitated them became, as the tremendous frenzy shook them more and more, more convulsive and more frenzied. After a few minutes of this, the female stood up. With her neck outstretched, her whole body shining and gleaming, she seemed to be offering herself. And with a slow effort, falling back several times as if his legs were powerless to lift his gigantic weight, the male reared up, resting his eager paws on the female's taut rump. And with roars of triumph and ferocious delight, the sauropods mated.[15]

The most remarkable feature of this description is that Mysor's sauropods are terrestrial and not aquatic, as was widely believed at the time.

For the sake of completeness, I won't spare you a colorful literary genre that has emerged in recent years: dinosaur erotica. As the name suggests, this is a subgenre of monster erotica that depicts sexual encounters between humans and various monsters. In dinosaur erotica, the sexual encounters are between humans and dinosaurs of the opposite sex or, in a gay variation, the same sex. A noteworthy author of this little-known genre is American Christie Sims, whose books discuss the reproductive organs of various dinosaur species, from *Velociraptor* to *Triceratops* and, of course, *Tyrannosaurus*. The titles sum up these pornographic stories: *Ravished by the Triceratops*, *Taken by the T-Rex*, and so on. Pterodactyls and pachycephalosaurs also appear, so there's something for everyone in these stories. But, let's be honest—they're extremely bad. The plot is repeated ad infinitum, with a vaguely prehistoric young huntress who finds herself the captive of a male dinosaur, who violates her. We might have hoped for some imagination with respect to the reproductive organs of these rapist dinosaurs, but alas, as described by the authors (and especially the female authors) of dinosaur erotica, the dinosaurian penis bears a (very) close resemblance to its human counterpart. It's a shame, of course, but at the same time we didn't expect Ms. Sims and her colleagues to write a thesis on the reproductive biology of dromaeosaurids and the evolution of the penis in archosaurs.

This American genre recently arrived in France with Grom's *T-rex mon amour* (My love *T. rex*).[16] I don't mean to be overly patriotic, but the French dinosaur erotica is a few leagues ahead of its American model. At least the author has done more research into dinosaur sexuality than her American counterparts, mentioning a spiral penis for the *T. rex* (like the duck's, that is), while the rest of the story is difficult to summarize and completely off the wall.

After Love

Once mating was complete, it was time to prepare the nest and lay the eggs.

It seems all dinosaurs laid eggs. At the end of the 20th century, some paleontologists suggested the possibility that sauropods were ovoviviparous (incubation and hatching of eggs occurred inside the mother's womb, and she then gave birth to living young), a hypothesis elegantly ruled out by the discovery of sauropod eggs. That's the charm of paleontology: beautiful, seductive hypotheses are destroyed in an instant by an ugly little fact. It's perfectly safe to hypothesize that sauropods didn't lay eggs, but when someone brings you back an egg with a sauropod embryo inside it, you just have to pack it in. Amniotes have two oviducts, elegantly named "fallopian tubes" in mammals, except birds, which only have one. This is why hens lay their eggs one at a time. Crocodilians have two functional oviducts and can lay several dozen eggs, either in a hole dug in the sand or in a nest made of soil and plant debris. Dinosaurs also had two functional oviducts, as shown by the discovery of two eggs inside the skeleton of a *Sinosauropteryx*. And, like crocodilians, they probably laid many eggs at the same time. At least that's the story their nests seem to tell us because, of course, we know of many fossilized dinosaur nests.

In the South of France, dinosaur eggs are a fairly common sight from Provence to Languedoc. An engineer from Marseille and a clergyman from Ariège were the first to encounter them in the mid-19th century.[17] The man from Marseille was Philippe Matheron, and the man from Ariège was Jean-Jacques Pouech, a canon, teacher at the seminary, and great geologist. It was in Le Mas-d'Azil, a township in the Ariège department, that Pouech discovered curious small calcareous fragments. He pondered over their characteristics for a long time, imagining pieces of eggshells from giant birds. As for Matheron, he didn't discover shell fragments that were difficult to interpret but rather two large half spheres in which he thought he recognized the eggs and remains of *Hypselosaurus*, a gigantic crocodilian. Since then, *Hypselosaurus* has had its trials and tribulations, first being reinterpreted as a sauropod dinosaur before being consigned to the dustbin of paleontology by paleontologists who considered the remains too incomplete to determine a species. But even if *Hypselosaurus* no longer exists, the eggs discovered by Philippe Matheron remain and are likely titanosaur eggs.

Since these first discoveries, hundreds of eggs have been found in these same geological formations dating from the end of the Cretaceous Period. The main thing these eggs from the South of France tell us is that dozens of nests were built next to each other and on several stratigraphic levels, generally belonging to titanosaurs, the large herbivores of the time. This information has been interpreted to mean these dinosaurs gathered to lay their eggs in the same place and returned year after year to the same egg-laying sites. This type of nesting colony, well known in birds and crocodilians, certainly makes it easier to guard the nests against the countless predators always on the lookout for a good omelet. These possible consumers included small mammals, lizards, and of course a whole host of carnivorous dinosaur species!

According to a group of researchers from Institut Català de Paleontologia in Barcelona, Spain, who have skillfully tackled this complicated subject, titanosaur clutches were elongated, bowl shaped, and measured 2.5 meters (8.2 feet) long.[18] They resembled a gravy boat, and since I probably haven't seen one on a table since the 1970s, I'm aware that my youngest readers may not have any image in their heads when they read this word. A gravy boat "is a low jug or pitcher with a handle in which sauce or gravy is served," and its "typical shape is considered boat-like, hence the name."[19] According to these researchers, these gravy-boat-shaped nests were dug by the females with one of their hind legs, and they contained an average of twenty-five eggs. The asymmetrical shape of a titanosaur foot with its inner toes bearing the largest claws perfectly explains the asymmetry of the nests, which are deeper on one side. With just a kick or two, the female would dig her gravy boat right into the ground and all that was left to do was release the eggs.

The mechanics of egg laying itself raises a few questions when we think of the largest dinosaurs. Take, for instance, the giant titanosaurs whose cloacae were 5 or 6 meters (almost 16.5 to 20 feet) above the ground. Did the female squat down or drop her precious eggs from that height, or did she have an ovipositor (a specialized organ for depositing eggs) of some kind? Did she build a large pile of vegetation beforehand on which to lay her eggs? We don't know, but in any case, after they were laid, these eggs were then covered with sediment or plants to maintain the correct incubation temperature. It seems some titanosaurs from the area that is now Argentina even laid their eggs in volcanic environments to take advantage of the proximity of hot springs to incubate their eggs. Which was pretty smart for the bottom of the class, which I'll remind you our poor sauropods were.

After more than a century and a half of collecting dinosaur eggs, we're just beginning to get an idea of their diversity. Following the French pioneers' work in the mid-19th century, the discovery of numerous eggs during the American Museum of Natural History's expeditions to Mongolia in the early 1920s generated euphoric stories in the press around the world, with headlines like "the first discovery of dinosaur eggs." This international reporting was a spectacular failure because it totally forgot the first French discoverers, Philippe Matheron and Abbé Pouech. The eggs found in Mongolia were studied by Henry Fairfield Osborn, president of the museum and the initiator of the great expeditions led by Roy Chapman Andrews, another interesting character. In 1923, Andrews enthusiastically recounted the discovery of these "first dinosaur eggs" at the Flaming Cliffs in southern Mongolia. They made the front page of *L'Illustration* magazine, a weekly illustrated French news periodical, on Saturday, December 22, 1923. Unexpectedly, they became the most remarkable and talked-about discovery of the museum's expeditions to Mongolia. Unexpected because Andrews and his team hadn't gone looking for dinosaurs in the first place, but rather had been searching for fossils of primitive humans, which Osborn had imagined would be found in this part of the world that at this time was considered the cradle of humanity. These expeditions from Beijing to the heart of the Gobi Desert, which Andrews recounted in his 1926 book *On the Trail of Ancient Man*, were utter failures. The museum's field team overcame innumerable difficulties, from attacks by Mongolian bandits to shortages of burlap to make cases (out of flour paste) around the discovered bones. Nothing was off limits to make up for the shortages, recounted Andrews. First the not-so-essential parts of the tents were cut away, then the team's clothes, even Andrews's own pajamas.

BANTER BETWEEN LOVERS

Andrews confessed to one regret: that these first dinosaur eggs ever found (well, he reluctantly admitted that perhaps a few fragments had also been found in France) measured only 20 to 23 centimeters (8 to 9 inches). The public was terribly disappointed by this modest—even ridiculous—size for dinosaurs, he confided. The few eggs that had appeared in literature before Andrews and his pals discovered them for real looked quite different. In one of the earliest dinosaur short stories, Robert Duncan Milne's "The Iguanodon's Egg," a gigantic *Iguanodon* spread terror in the 19th century after hatching from an egg on December 22 on a lost island off the coast of Papua New Guinea.[20] This egg from which *Iguanodon* emerged was of truly dinosaurian proportions— 4.3 meters (14 feet) in diameter! Even though the creature grew to be 33 meters (100 feet), a 4.3-meter or 14-foot in diameter egg was still huge! So by comparison, the 20- to 23-centimeter or 8- to 9-inch eggs found in Mongolia really were a pittance. And the public was disappointed.

Osborn boldly concluded that these eggs, found alongside skeletons of a curious toothless theropod dinosaur, must have belonged to a small marginocephalian abundant in the deposit. He named the marginocephalian *Protoceratops*. The toothless animal found fossilized with the *Protoceratops* eggs must have been an animal that ate eggs. Logically, Osborn named it *Oviraptor philoceratops* (meaning "egg thief"). When, at the end of the 20th century, it was discovered that these *Protoceratops* eggs actually contained *Oviraptor* embryos, one of paleontology's most cherished legends collapsed: we no longer knew what *Oviraptor* ate, and *Protoceratops* eggs were no longer known to exist. But since the rules of zoological nomenclature are extremely strict, it's impossible to change the now ill-suited name of *Oviraptor*.

A few years later, true *Protoceratops* embryos were discovered, again by teams from the American Museum of Natural History, who one hundred years after their expeditions to the Gobi Desert continued to make remarkable discoveries.[21] Reputable museums (and the American Museum of Natural History was the home of our friend Barnum Brown) sometimes keep their reputable habits. And a big surprise was that these *Protoceratops* eggs were soft shelled! A big surprise because crocodilians like birds lay hard-shelled eggs, so all dinosaurs were expected to have laid hard-shelled eggs too. The museum's researchers therefore hypothesized that the first dinosaur eggs would have been soft shelled and that mineralized shells would have appeared several times during the group's evolution, thus explaining why eggshells were never discovered from before the Late Jurassic! It also explains why calcareous eggshells have been discovered for only a few groups of dinosaurs: titanosaurs, hadrosaurs, and maniraptorans (the branch of theropods that includes birds). Shell calcification is thought to have occurred three times in dinosaur evolution, but most species would have laid soft-shelled eggs.

Heyuannia's Blue Eggs

Hard-shelled eggs are always white when laid by crocodilians or turtles, while colored eggs seem to be characteristic of birds. Like all supposed bird characteristics, such as feathers, the question has been raised as to whether this characteristic might have appeared earlier in the evolution of the dinosaur group. To determine the color of dinosaur eggs, which is not visually preserved, researchers have used Raman microspectroscopy, an exceptional tool that also equips the *Perseverance* rover on Mars. But before heading off to Mars to do I don't know what, Raman microspectroscopy proved its worth on dinosaur eggs.[22] The results were intriguing: the ornithopod and sauropod eggs studied showed no traces of pigment and therefore should logically have been white like crocodilian eggs. Maniraptoran eggs, on the other hand, contain protoporphyrin (a red-brown pigment) and biliverdin (a blue-green pigment) in variable quantities. The proportion of these two pigments, responsible for the color of bird eggs, led to the assertion that the oviraptorid dinosaur *Heyuannia* laid blue-green eggs like emus and like the raptor *Deinonychus*. Troodontids had yellow-orange eggs like the majority of chicken eggs (which are loaded with protoporphyrin). Many dinosaur eggs were also speckled and spotted, like many modern bird eggs. This is generally considered a camouflage strategy—yet another strategy—adopted by dinosaurs before they became birds. Colored eggs therefore predate the first birds. So far, the oviraptorid *Heyuannia* is the most primitive theropod whose eggs have been studied, but no allosaur eggs or those of other non-maniraptoran theropods have been examined, so it's possible that these pigments appeared earlier in theropod evolution. It's also possible that some ornithischian or sauropod eggs were colored, but the few shell samples studied so far have definitely been white.

Oviraptorids from Mongolia laid their eggs in the open air, and the parents incubated them the way the first ostriches did. Several skeletons of oviraptorids have been discovered in Mongolia in an incubating position, confirming that some carnivorous dinosaurs behaved exactly like birds. Oviraptorid eggs are generally elongated, with one end thinner than the other, and were laid in pairs (meaning mothers had two functional oviducts) in three successive circles. The mother would lay the first circle of eggs almost vertically, circling around a mound of vegetation prepared beforehand; she would cover the freshly laid eggs with sediment, leaving their rounded tops protruding, then move on to the second circle and then the third. In total, an oviraptorid nest could contain thirty-two eggs. There was always an empty space in the middle of the nest, which could have been occupied by an adult. This adult would have been covered in large feathers, and it was likely incubating the nest regularly.

In addition to *Oviraptor*, several other oviraptorid genera such as *Citipati*, *Khaan*, and *Nemegtomaia* built this type of nest and were able to incubate their eggs there. The giant oviraptorids (*Gigantoraptor* reached 8 to 9 meters [26.2 to 29.5 feet] in length and weighed over 2 metric tons) must have done the same as their smaller cousins, since eggs measuring 43 centimeters (1.4 feet) in length have been found in an arrangement exactly like *Oviraptor*'s. Other large eggs measuring 45 centimeters (1.5 feet) long were found with an associated *Beibeilong* embryo, a genus of large caenagnathid. Caenagnathidae is a family closely related to Oviraptoridae within the Oviraptorosauria group. As with many Chinese fossils, the story of *Beibeilong* (meaning "baby dragon") is a rather muddled one, beginning with illegal excavations, an under-the-counter sale in the United States, and then repatriation to the Celestial Empire. What is certain is that these eggs, with a volume of around 5 liters (1.3 gallons), are the largest non-avian dinosaur eggs known today. The world-record holder for the largest egg is *Aepyornis*, the elephant bird of Madagascar, which has denied all dinosaurian attempts to dethrone it. It's one of the largest birds ever to have existed, with a height of 3.5 meters (11.5 feet) and weighing 350 to 500 kilograms (around 770 to 1,100 pounds); its close cousin *Vorombe* reached 650 kilograms (over 1,400 pounds). *Aepyornis* became extinct between around 1000 CE and the 18th century, and laid eggs measuring 1 meter (3.3 feet) in circumference, 35 centimeters (1.2 feet) in diameter, and 9 liters (2.4 gallons) in volume. A record still to be beaten (the capacity of an ostrich egg is 1 to 2 liters [0.3 to 0.5 gallons]).

Troodon is a close cousin of *Velociraptor* and *Deinonychus*. Osborn's error was not the last of its kind, as the nests of this small theropod from North America were first mistakenly identified as those of the small ornithopod *Orodromeus*. These nests contained up to twenty-four elongated, asymmetrical eggs laid almost vertically in an 80-centimeter (2.6-foot) circle, with the thinner end pointing downward, like in oviraptorids. Numerous *Orodromeus* bones of various sizes were found around the nests, hence the original identification of the eggs belonging to them. The subsequent discovery of embryos in the eggs, however, as in the case of *Oviraptor* and *Protoceratops*, changed the game: *Troodon* embryos were found in the supposed *Orodromeus* eggs, which may well have been the *Troodon* parents' prey! It was a major identification error, but before the embryos were discovered, the hypothesis wasn't absurd.[23]

Good Mothers

Ornithopods are the last major dinosaur group whose eggs, embryos, and nests are known. The best known is certainly *Maiasaura*, the "good mother lizard" described by American paleontologist Jack Horner in the late 1970s. In addition to the "*Orodromeus* nests" that became *Troodon* nests, Horner discovered about fifteen skeletons of juvenile hadrosaurs in Montana, measuring about 1 meter (3.3 feet) long and fossilized on a sort of bed of eggshells. His interpretation was that these animals were juveniles that had already hatched, and not embryos, and that their presence in a nest implied parental care and nesting behavior. According to Horner, the *Maiasaura* juveniles spent enough time in the nest to shed their eggshells and double in size. They would have spent several months there, likely under the supervision of adults.

Hypacrosaurus is another hadrosaur from North America, discovered and named by Barnum Brown in 1913. It has a rounded crest not unlike *Corythosaurus*'s crest. The adults approached

10 meters (32.8 feet) long and weighed 4 metric tons. *Hypacrosaurus* eggs are spherical and around 20 centimeters (almost 8 inches) in diameter, with embryos that when laid out can measure 60 centimeters (23.6 inches) from the tip of their adorable little beaks to the end of their tails. It seems that, unlike *Maiasaura*, young *Hypacrosaurus* left the nest immediately after hatching. So, within the same family, there are nidicolous species (*Maiasaura*) and nidifugous species (*Hypacrosaurus*), which makes us cautious about overgeneralizing.

Of course not all dinosaurs incubated their eggs. For physical reasons alone, it's hard to imagine a *Diplodocus* sprawled over its nest to keep it warm! Many species, especially sauropods, probably buried their eggs in incubation mounds—piles of vegetation prepared in advance, which, by fermenting over a period of months, would maintain a stable temperature.

Life in the Egg

The incubation period of dinosaur eggs seems to be one of those evanescent things from the past that we'll never fully know. Incubation time doesn't fossilize, does it? Yet we have a way of quantifying this duration, thanks to the work of Victor von Ebner (1842–1925), a distinguished Austrian anatomist. From his long scientific career, posterity has particularly remembered the description of growth lines in the dentine of teeth, now known as "von Ebner lines." These lines correspond to an increase in mineralization during the day, so each line represents one day. By counting the von Ebner lines in a tooth, we can theoretically tell how many days it took to form. These lines are found in the teeth of mammals, crocodilians, and dinosaurs. Growth lines have also been noted in the teeth of crocodilian embryos. Tooth formation (and thus the deposition of von Ebner lines) begins during embryonic development; in crocodilians, this starts at between 42% and 52% of the incubation period. In other words, the total incubation period corresponds roughly (in days) to a little more than twice the number of von Ebner lines counted on a crocodilian tooth at the time of hatching.

Incubation lasts 80 to 90 days in crocodilian species and 21 days in chickens. But how long was it for dinosaurs? American researchers have provided the answer for two species of ornithischian dinosaurs: *Protoceratops andrewsi*, a protoceratopsid from Mongolia, and *Hypacrosaurus stebingeri*, a hadrosaurid from Canada.[24] Two embryos were discovered inside their eggs, and the embryonic teeth of these little monsters had von Ebner lines! Given their stage of development, both dinosaurs are thought to have died shortly before hatching. These paleontologists counted forty-eight von Ebner lines in a *Protoceratops* tooth, so assuming the tooth began to form at 42% of the total incubation time (as in crocodilians), this equates to a total incubation duration of 83 days for a *Protoceratops* egg. Things were a little more complicated in the *Hypacrosaurus* embryo because hadrosaurs' teeth were stacked in structures called "dental batteries," with three teeth in each socket. The oldest tooth was already worn down to the root (embryos used to grind their teeth; just imagine the racket when approaching nests), and its growth lines could no longer be counted. That left the two most recent teeth, and by counting their respective von Ebner lines, it was determined that the time elapsed between the start of each tooth's formation (tooth replacement rate) was 44 days. Since the second tooth was 55 days old, it would also have started forming 44 days after the first tooth (which was just a root remnant) began developing. This gives a total odontogenesis (tooth formation) duration of 99 days. Since odontogenesis starts at around 42% of the incubation time, this gives *Hypacrosaurus* a minimum incubation period of 171 days, which equates to almost six months from egg laying to hatching!

Since that study, the same method has been used to estimate that the incubation period of the small theropod *Troodon* was 74 days.[25] Theropods therefore appear to have had an intermediate incubation period between the comparatively shorter incubation period of birds and longer one of crocodilians. In any case, there was a lot of gnashing of teeth inside of dinosaur eggs. Were these the beginnings of communication between the fetuses? Did they start emitting sounds in the days leading up to hatching? Did their parents communicate with them while still inside their eggs, as crocodiles seem to?[26] Even before hatching, young Nile crocodiles start to produce calls inside their eggs, emitting high-pitched sounds, which seems to lead to the hatching of other eggs and the arrival of the mother, who opens the nest, helps the late hatchlings to hatch, and gently carries the newborns into the water. It's quite plausible that some dinosaur species did this too.

Among the latest discoveries on the subject of hatching comes the work of two research teams who have recently concluded that oviraptorid egg hatching is asynchronous. In crocodilians and other reptiles, all the eggs in the same nest hatch at the same time, which is quite normal since they are also all laid at the same time. But a study of the developmental stage of different embryos in the same *Oviraptor* nest showed that some were much closer to hatching than others.[27] Hatching asynchrony is rare in modern birds, and its presence in a dinosaur only confirms once again that birds didn't invent much. Still, in some dinosaur species, all young were not born together, but one after the other, which clearly suggests that all eggs were not laid in a single day but over a longer period.

Paleontologists have described an egg tooth in a titanosaur from Latin America. An egg tooth is a small hornlike structure at the end of the beak or snout of oviparous animals, which helps hatchlings to break the shell that separates them from the outside world. If the egg tooth wasn't enough to break through, they could always emit cries for parental assistance. Remember the frequency of hadrosaur cries at birth was within the hearing range of their parents, so they could announce the good news of their hatching. Also remember that some dinosaurs' jaws, such as tyrannosaurs', were equipped with numerous ISOs (integumentary sensory organs), giving them an exceptional sense of touch. This may have enabled them to help hatchlings or transport them to crèches, similar to crocodiles. At any rate, after hatching we have seen that some species remain in the nest for a while, likely being fed by their parents. This was true of the hadrosaur *Maiasaura*, whose crèches have been discovered. Many other hatchlings were left to fend for themselves as soon as they hatched, embarking on a journey to adulthood punctuated by difficulties few could overcome and many succumbed to. Most of the young dinosaurs, however, could have been singing the song "You'll Never Walk Alone" as they set off to discover the dangerous outside world with their many siblings. And what did the youngsters do in their youth, in the company of their siblings? Play, of course!

Did *T. rex* Play? An Outlandish Hypothesis

Many modern animal species play. Ethologist Gordon M. Burghardt has devoted a five-hundred-page book, *The Genesis of Animal Play*, to the subject, inventorying the existence of play in zoological groups as diverse as insects, mollusks (cephalopods but, rest assured, not oysters), crustaceans, and of course vertebrates.[28] According to Burghardt, the five criteria required to define play in an animal species are that the behavior is not for the animal's current survival, is done for its own sake, differs from ethotypic behavior, is performed repeatedly, and is initiated when the animal is well fed, healthy, and unstressed.[29]

WHAT DID DINOSAURS THINK ABOUT?

As play is common in crocodilians and birds, its practice in dinosaurs seems certain for phylogenetic reasons already mentioned many times. So let's get one thing straight about crocodilian play: as with tools, don't expect anything too elaborate, or you'll be disappointed. Crocodilians don't play chess or water polo, but at least in captivity, they do like to play with the floating objects given to them by zookeepers. In crocodile crèches, young individuals are sometimes carried on the backs of slightly older individuals. Crocodiles also play with food. A Nile crocodile was observed throwing a dead young hippopotamus into the air over the course of 25 minutes. To sum it up, crocodiles are jokesters, and it's entirely plausible that dinosaurs may have engaged in play. In Michael Crichton's *The Lost World*, the sequel to *Jurassic Park*, baby tyrannosaurs also play with food. Their meal on this occasion was the villain Lewis Dodgson, whom they ate alive—but he had it coming! But the real world is even more violent than the movies. American paleontologist Bruce M. Rothschild of the University of Kansas reported the discovery of several *Triceratops* bones, especially occipital condyles, with traces of tyrannosaur bite marks.[30] The occipital condyle is a round prominence of bone attached to the base of the skull and articulates with the first neck vertebra. To get your hands on such an object, you'd have to rip off a *Triceratops*'s head, turn it upside down, and then sever the condyle. This probably wasn't done for gastronomic reasons, since the meat of the *Triceratops* was located elsewhere, in its limbs, torso, and abdomen. We also suspect this wasn't done to consume the animal's tiny brain. So in his paper, Rothschild seeks to answer the question: why bother decapitating a *Triceratops*? According to him, there's only one explanation, and that's to retrieve this spherical object measuring about 4 inches in diameter and then play ball with it! This may make you smile, or even shrug, but Rothschild is a serious specialist in dinosaur paleopathology and a forensic paleontologist who shares Sherlock

Holmes's motto: when you have excluded the impossible, whatever remains, however improbable, must be the truth. So maybe tyrannosaurs had fun throwing their little bone balls into the air, or maybe they used them as stress balls—who knows!

Painful Lessons

When an immature juvenile tyrannosaurid specimen was found with a deformed skull caused by severe bites, it led to another hypothesis about learning behavior.[31] The injured dinosaur's snout bones were pierced by four elliptical teeth—most likely the teeth of a fellow tyrannosaurid of the same size. The attacker must have taken the victim's entire upper jaw in its mouth and squeezed hard enough to pierce the skull bones but not hard enough to tear them out, an argument for learning behavior that became a little too aggressive. The attacked animal survived the aggression but died young, probably around the age of twelve to fourteen years old. The fight took place when the two similarly sized tyrannosaurs were not yet sexually mature, which rules out the hypothesis of sexual competition or conflict, which is common among adults. After ruling this out, the only explanation really left for this violent behavior is that it was a learning behavior, imitating (in this case, violently imitating) the conflict competing adults engaged in over mates, food, or territory. Modern crocodilians frequently suffer injuries like these related to turf wars, especially in the case of the saltwater crocodile (*Crocodylus porosus*), which is particularly attached to its territory and fiercely defends it against intruders.

EPILOGUE

Final Thoughts

All evidence to date proves dinosaurs were not the cold-blooded, solitary, unintelligent animals they've long been portrayed to be. At the end of our exploration of their lives, we're left with a few certainties: They were warm-blooded, most were social animals, they felt pain, and they likely helped each other. They perhaps even loved playing! In a nutshell, the idea Marcellin Boule peddled in 1905 about dinosaurs (they "were unintelligent beings") simply isn't true.

Steven Spielberg and Michael Crichton's images of highly intelligent raptors hunting in packs and of tyrannosaurs catching a jeep at full speed, however, need to be toned down a little in light of the latest studies. Even though *T. rex* had good acceleration, it didn't have endurance. While dromaeosaurs—the raptors—were undeniably intelligent, their intellectual capacities appear to have been far inferior to crows'.

What will be the next great advances in our understanding of dinosaur behavior? They're waiting in boxes—boxes full of fossils—to be analyzed by a CT scanner or microspectrometer or some other future instrument. They're also dependent on whatever new, unpredictable discoveries are made, such as finding a tyrannosaur nest, an entire psittacosaur cloaca, a sauropod mummy, or the nasal horn of an *Ornitholestes*. Rest assured, paleontologists are certainly at work with chisels and pickaxes somewhere in the world as you're

EPILOGUE: FINAL THOUGHTS

reading this book, and many amazing discoveries are being exhumed and prepared before they're announced. The discoveries of recent years also show that we must be wary of unjustified phylogenetic interpretations. This means that a trait present in crocodilians and birds will not necessarily be present in all dinosaurs. For example, eggs with calcareous shells were not inherited from a common ancestor of birds and crocodilians but evolved convergently on several occasions.

Molecular paleontology is one of the most promising areas of study for revolutionizing our understanding of dinosaurs. Not so long ago, paleomolecules were thought to have disappeared forever during the process of fossilization. Since the beginning of the 21st century, a number of studies have proven this belief to be false. We've seen that the colors in eggs and skin have left molecular traces in the form of biliverdin or melanin. It's clear we'll be pursuing this field of study in the coming decades. Will we be able to resurrect dinosaurs using small bits of DNA and fragments of protein, leaving behind the shortcomings of paleoethology for the rigor of ethology? Less than ten years ago, it was thought that DNA could not survive more than a few tens of thousands of years, then it was hundreds of thousands of years, and at the end of 2021, the oldest DNA passed the million-year mark![1] It was discovered in 1.1-to-1.6-million-year-old mammoth teeth extracted from permafrost in Russia. The tooth fragments were degraded, but considerable progress in sequencing made it possible to analyze them. At this exponential rate, in a matter of 10 years from now, 100-million-year-old DNA will be showing up! Therefore, it would be rash to assert we'll never find bits of dinosaur DNA, even if researchers believe the age limit for ancient DNA is the same as permafrost, which is 2.6 million years and clearly insufficiently old enough to contain dinosaur DNA. The oldest Antarctic ice dating from the Oligocene Epoch (around 34 million years ago) is unlikely to contain the remains of animals that disappeared 66 million years ago.

As we conclude this tour through the brains of dinosaurs, there are far more questions than answers. In any case, the de-extinction of dinosaurs is not in sight for the foreseeable future, which saves me from being challenged too soon by ethologists studying *Tyrannosaurus* games and *Triceratops* fights, this time through binoculars, or by others flying drones over the wavy fronts of *Pachyrhinosaurus* herds or chasing *Gallimimus* herds to measure the top speed of this agile dinosaur.

APPENDIX

The Dinosaur Family Tree

Without going into details that are of little interest to me (and therefore probably to you), I thought it would be useful to give a general overview of the dinosaur classification discussed throughout these pages. We'll look at the major phylogenetic hypotheses concerning the various groups, bearing in mind that these hypotheses are often highly volatile because building phylogenetic trees from incomplete skeletons, or even just a few bones, is risky. Some of the hypotheses presented here could easily be disproved tomorrow.

The Very First Dinosaur

At the beginning of the Late Triassic Period 235 million years ago, a single landmass (Pangaea) was surrounded by a single ocean (Panthalassa). At the far southwest of this immense continent was present-day Argentina. In Ischigualasto Provincial Park in northern Argentina, Carnian geological layers were deposited in rivers and floodplains bordered by active volcanoes in a climate with heavy seasonal rainfall. Although the exact vegetation is not known, it was largely composed of horsetails, ferns, and large trees standing over 40 meters (more than 130 feet) tall, perhaps closely related to *Araucaria*.

APPENDIX

The big predator of the time was not a dinosaur but a poposauroid, another member of the clade Archosauria. *Sillosuchus*, a poposauroid archosaur, could reach 9 meters (nearly 30 feet) in length. Other big beasts that roamed Argentinian soil 230 million years ago included synapsids (the amniote group that includes mammals), such as *Ischigualastia*, a herbivorous dicynodont 3 to 4 meters (9.8 to 13.1 feet) long that weighed 1 to 2 metric tons. Finally, *Eoraptor* (1 meter [3.3 feet] long), *Eodromaeus* (1.2 meters [3.9 feet]), *Panphagia* (1.3 meters [4.3 feet]), and *Herrerasaurus* (4 to 5 meters [13.1 to 16.4 feet]) together made up around 11% of the Ischigualasto fauna[1] and are the oldest dinosaurs known to date. At the beginning of the Late Triassic Period, dinosaurs were only a marginal part of the terrestrial vertebrate populations, dominated by synapsids and other archosaur groups.

The oldest dinosaurs currently known date from the beginning of the Late Triassic Period, between 235 and 230 million years ago. Among them, *Saturnalia* and *Pampadromaeus* from what is now Brazil, and *Panphagia* from what is now Argentina are considered primitive sauropodomorphs (remember that sauropodomorphs include *Diplodocus*, its cousins, and its ancestors). Also from Argentina, *Herrerasaurus* and *Eoraptor* are classified as either sauropodomorphs or theropods. *Eodromaeus* is definitely a theropod. In any case, both groups of saurischians (theropods and sauropodomorphs) originated as early as the Late Triassic Period, whereas the oldest known ornithischians only date from the beginning of the Jurassic Period, 30 million years later. *Pisanosaurus*, another skeleton from the Ischigualasto Formation in Argentina, has long been considered the oldest ornithischian, but it was later reclassified as a silesaurid, which we'll discuss later. As for the other ornithischians supposedly from the Triassic Period, the geological formations in which they were discovered have been redated to the beginning of the Jurassic Period. Given that we only have fragments of this *Pisanosaurus* specimen, the possibility of a further change can't be ruled out. In short, as of 2022, there are no ornithischians from the Triassic Period. Absence of proof is not proof of absence, so no one would be shocked if a Triassic ornithischian appeared in the next few years.

The search for the oldest dinosaur regularly comes up against two stumbling blocks. The first is phylogenetic: an animal like *Pisanosaurus*, long considered the oldest ornithischian, finds itself kicked out of the dinosaur group when new research shows its characteristics better match those of another archosaur group. The second is geological: the continental formations in which dinosaur skeletons are found are sometimes loosely dated because of a lack of sufficient research. After more detailed geological studies, an "older" dinosaur may actually be tens of millions of years younger than originally thought. This happened to the former "oldest sauropod," *Isanosaurus*, a herbivorous dinosaur measuring around 10 meters (32.8 feet). When it was first described, the geological layer that contained the specimen was dated to the Norian Stage (Upper Triassic, 210 million years ago). A few years later, an in-depth study of the pollen grains in the geological formation in Thailand in which *Isanosaurus* had been discovered showed that, while the lower part of the formation indeed dated to the Norian, the upper part was Jurassic. Since *Isanosaurus* came from the top of the formation, out went the oldest sauropod title.[2]

The discovery of numerous primitive dinosaurs living in the Carnian Age suggests the group originated a few million years earlier, perhaps during the Ladinian Age (242–237 Ma) or even the Anisian Age (247–242 Ma), but in any case during the Middle Triassic Period. Several research teams on the trail of the oldest dinosaur in South American and African geological formations dating to this period identified a suspect, *Nyasasaurus*.[3] *Nyasasaurus* was discovered in the 1930s during major British expeditions to Tanzania, near Lake Nyasa, and was named in 2013 (even paleontologists sometimes need to let time take its course). At 243 million years old, this little animal from the Anisian Age (an age named after the river Enns in Austria, whose Latin name is *Anisus*) is potentially the oldest dinosaur. Unfortunately, *Nyasasaurus* is only known from a single humerus bone, which isn't much. But if it was indeed a dinosaur, this could push back the group's first appearance by a few million years to the Early Triassic Period, just after the great extinction event at the end of the Paleozoic Era, which caused the extinction of 80–90% of living species 251 million years ago.

The ancestor in question is a creature that must have looked a lot like the proto-dinosaur *Silesaurus*. *Silesaurus* lived 230 million years ago in Silesia, a region in southwestern modern-day Poland. It gave its name to a family of small Triassic archosaurs, the silesaurids, the closest group to the dinosaurs (the "sister group" in the jargon of phylogeneticists). Short of finding an older, more primitive dinosaur than those currently known, silesaurids probably give a pretty good idea of what the first dinosaur must have looked like. *Silesaurus*, known from complete skeletons, was a quadruped measuring a little over 2 meters (6.5 feet) long, including its meter-long (3.3-foot) tail, and around 70 centimeters (2.3 feet) high at the withers. It had vertical legs under its body like dinosaurs, a sort of horny half beak (on the lower jaw only), and a diet made up of plants and insects, according to the composition of coprolites attributed to it.[4]

Theropods

Theropods include all carnivorous dinosaurs, as well as a few families that are probably omnivorous or even vegetarian, and all present-day and fossilized birds. They are defined as a group of saurischians more closely related to birds than to sauropodomorphs. Alternatively, they can be defined as the most inclusive clade, which contains *Passer domesticus* (or any bird of your choice) but not

Saltasaurus loricatus (a sauropod from Argentina). If you think about it, these definitions mean exactly the same thing. Excluding birds, 350 species of theropod dinosaurs have been described, representing a remarkable amount of diversity and a considerable variety of forms both morphologically and ecologically, and ranging from huge predators to small herbivores. While some lost their teeth in the course of evolving toward less animal-heavy diets, the vast majority had characteristic knife-blade teeth with sharp crenulations. Almost all of them were digitigrade bipeds with three functional toes, hollow bones, and often reduced-size or even vestigial arms, such as in *Carnotaurus* and *Tyrannosaurus*.

The oldest currently known theropod is probably *Eodromaeus* (meaning "dawn runner" in Greek), which lived in the Ischigualasto-Villa Unión Basin 230 million years ago, toward the end of the Carnian Age (the first age of the Late Triassic Period). This little animal, known by a complete skeleton, measured 1.2 meters (nearly 4 feet) from snout to tail, 45 centimeters (nearly 1.5 feet) high at the withers, and weighed 5 kilograms (11 pounds) at the most. Its remains were found in the same geological formation (Ischigualasto Formation) as dinosaurs such as *Herrerasaurus* and *Eoraptor*, which are saurischians but probably not theropods, according to the latest phylogenetic studies. *Eodromaeus*'s teeth have a knife-edge morphology with small crenulations, characteristic of theropods.

Ceratosaurs are a branch of theropods that diverged early on, evolving right up to the Cretaceous-Paleogene boundary. It includes abelisaurids, large predators that scoured the southern continents and Europe during the Late Cretaceous Period. The scant data available suggests that ceratosaurs had scaly skin and no feathers. Some ceratosaurs had perplexing adaptations, like *Berthasaura* from the Cretaceous Period of Brazil who lost its teeth, or its cousin *Masiakasaurus* from Madagascar who had peculiar fan-shaped teeth. But the majority of ceratosaurs were predators, like *Carnotaurus*, an abelisaurid from Argentina with arms extremely reduced in size, which was an evolutionary convergence with *Tyrannosaurus*, who is not a ceratosaur but a tetanuran.

Tetanurans include carnosaurs and coelurosaurs. Some carnosaurs—the megalosauroids—seem to have been irresistibly attracted to waterways and their inhabitants. Among the megalosauroids were the spinosaurids *Baryonyx* and *Spinosaurus*, which had long snouts and conical teeth, a remarkable evolutionary convergence with present-day crocodilians. The large claw on *Baryonyx*'s hand is suggestive of a behavior analogous to the grizzly bear behavior of using its claw to catch salmon in the water and throw them back on shore before feasting on them. The strange *Spinosaurus* from what is now North Africa may have been amphibious, but its cousins were at minimum completely dependent on waterways for their diet of fish. The other megalosauroid family, the megalosaurids, seems to have been equally fond of fish. Megalosaurids would have been great at cleaning the foreshore, roaming the beaches in groups in search of stranded prey. At least that's what recent research suggests at footprint sites in southern Portugal where hundreds of megalosaurid footprints intersect.[5] But living by the sea doesn't necessarily mean eating its fruits. And the best evidence of a dinosaur's diet is always provided by finding the remains of its last meal. Although long neglected, this evidence of the megalosaur appetite for fish may have been around for several decades. In fact, in 1838, when Jacques-Amand Eudes-Deslongchamps described the dinosaur *Poekilopleuron bucklandi* found in Normandy, France, another megalosaur from the Bathonian Age, he noted the presence of shark teeth (to be precise: a dozen teeth of *Polyacrodus*, a shark of the hybodont group). At this point, we might also think that *Polyacrodus* had pecked away at the dinosaur's carrion, leaving a few teeth behind, but Eudes-Deslongchamps also reported cartilaginous fragments in the rib cage, interpreting it to be the *Poekilopleuron*'s last meal, the plausible remains of a feast of cartilaginous fish.[6] Unfortunately, this can't be verified today, as the *Poekilopleuron* skeleton was destroyed along with its intestinal contents and the rest of the collections of the Caen Faculty of Science during a World War II bombing raid in 1944.

The other carnosaurs—allosauroids—are not known to have a comparable diet. These large predators, divided into several families, lived on every continent from the Middle Jurassic Period to the beginning of the Late Cretaceous Period. *Allosaurus*, from the Upper Jurassic of the United States, is known from numerous skeletons discovered in the same quarry, suggesting the possibility of sociality. A family of allosauroids from Africa and South America—carcharodontosaurids—includes some Cretaceous giants that rival tyrannosaurs in size, such as *Carcharodontosaurus* from the Sahara and *Giganotosaurus* from Argentina (measuring 12 to 13 meters [39.4 to 42.7 feet] long and weighing 6 to 8 metric tons).

Coelurosaurs, a sister group to carnosaurs, are defined as the clade comprising all theropods more closely related to birds than to carnosaurs. They include tyrannosauroids, ornithomimosaurs, and the extensive branch of maniraptorans.

The tyrannosauroids comprise many Cretaceous predators, including of course the biggest of them all, *Tyrannosaurus rex*, which lived in North America at the very end of the Mesozoic Era. The tyrannosaurids at the end of the Cretaceous Period were a remarkable bunch of bullies, measuring 8 to 14 meters (about 26 to 46 feet) long and weighing 4 to 10 metric tons.

As their name suggests, ornithomimosaurs looked like birds running around with toothless jaws and a horny beak. Ever since O. C. Marsh named the first ornithomimosaur discovered in 1890 *Ornithomimus* (meaning "bird mimic"), subsequent discoverers of this type of dinosaur have exploited the "mimus" affix to the full:

APPENDIX

Struthiomimus ("ostrich mimic"), *Gallimimus* ("chicken mimic"), *Dromiceiomimus* ("emu mimic"), *Anserimimus* ("goose mimic"), *Tototlmimus* ("bird mimic," the prefix is the Nahuatl word for "bird"), *Sinornithomimus* ("Chinese bird mimic"), and the mythological *Kinnareemimus* ("Kinnaree mimic"; kinnari are half-woman, half-bird beings in Southeast Asian myths), and *Harpymimus* ("Harpy mimic," in reference to the Harpies of Greek mythology, other half-woman, half-bird creatures). A variant on the same theme is *Rativates* ("ratite foreteller"; ratites are large flightless birds). Ornithomimosaurs are one of many theropod branches from the Cretaceous Period and are generally considered omnivores with a preference for vegetation. They all had long arms, which is particularly evident in *Deinocheirus* (meaning "horrible hand," an etymological exception due to the fact that its discoverers, the Polish paleontologists Halska Osmolska and Ewa Roniewicz, thought they had found another type of dinosaur), which for a long time was only known by its two 2-meter-long (6.6-foot-long) arms. After half a century of uncertainty about *Deinocheirus*, the discovery of complete skeletons in 2014 proved it had been an 11-meter-long (36-foot-long) ornithomimosaur with a duck-like bill, definitely a silly-looking animal.[7]

Maniraptorans are the large group of theropods that includes birds and all the related non-avian dinosaur families, including alvarezsaurids, therizinosaurs, oviraptorosaurs, scansoriopterygids, dromaeosaurids, and troodontids. All of them were covered in feathers and were the most intelligent of the group. Dromaeosaurids were the famous raptors from *Jurassic Park*, which were named after the genus *Velociraptor* from the Upper Cretaceous of Mongolia. *Velociraptor* was a small predator that measured 1.5 meters (almost 5 feet) long and 0.5 meters (1.6 feet) high at the hips, and weighed around 15 kilograms (33 pounds)—roughly the size of a turkey and much smaller than those Hollywood raptors! But among its many cousins, *Utahraptor* (from the Lower Cretaceous of the United States) reached 7 meters (23 feet) in length, and *Deinonychus* more than 3 meters (9.8 feet); the latter is closest to Spielberg's raptors, and quite rightly so, since it was his model. The author of the book *Jurassic Park*, Michael Crichton, spoke at length with the researcher who described *Deinonychus*, paleontologist John Ostrom of the Yale Peabody Museum of Natural History. But as he found the name *Velociraptor* catchier than *Deinonychus*, Crichton decided to use it instead.[8] The raptors in *Jurassic Park* (who came close to being called "nychus," which would have filled only the Hellenists with fear) are unquestionably *Deinonychus*, however. And like all maniraptorans, the real *Deinonychus* were covered in feathers, a fact that Crichton, Ostrom, and Spielberg didn't know when the first film was released in 1993. It was only at the very end of the 20th century that these dinosaurs were confirmed to have had feathers.

Among the other maniraptoran families, troodontids are tiny animals closely related to dromaeosaurids, with the same enlarged claws on their feet. Therizinosaurids are strange dinosaurs with gigantic hand claws (0.7 to 1 meter [2.3 to 3.3 feet] long!). These astonishing appendages would have enabled them to cut the branches on which they fed, or even to rip open termite mounds if they had a craving for protein supplements.

Their oviraptorid cousins were completely toothless but covered in feathers, and they're assumed to have been omnivores. Alvarezsaurids and scansoriopterygids represent other surprising evolutions of maniraptorans discovered recently. Alvarezsaurids were small bipeds with a single large finger on each hand. Scansoriopterygids were tiny animals measuring 15 to 25 centimeters (6 to 9.8 inches) long with one extremely long finger on each hand that was longer than their body. The long finger supported a membrane, probably enabling them to glide from branch to branch. These two families weren't described until the 21st century. Their discovery continues to unveil the extraordinary biodiversity of dinosaurs and confirms that many surprises still await paleontologists.

Sauropodomorphs

Sauropodomorphs are made up of the large sauropod group—the long-necked quadruped giants—and many other more primitive species. *Panphagia*, one of the inhabitants of the Ischigualasto–Villa Unión Basin 230 million years ago, is currently the oldest known sauropodomorph. Far smaller than its famous descendants, it measured 1.3 meters (4.3 feet) and weighed 4 to 5 kilograms (8.8 to 11 pounds). As its name suggests, *Panphagia* "ate everything," or at least ate anything. It was omnivorous, as the shape of its teeth suggests. Its contemporary, *Saturnalia*, was discovered in southern Brazil during the carnival season, which gave it its name, in reference to the Roman festival Saturnalia. It's also a very small creature. Dating of the Brazilian formation estimates an age of 233 million years, corresponding to the middle of the Carnian Age (237–227 Ma), in which case, *Saturnalia* would be between 231 and 228 million years old, just a little older than *Panphagia*. This is also the age of *Buriolestes*, a third primitive sauropodomorph also from the area that is now Brazil. Its osteological and cerebral characteristics show it to have been a small carnivore, confirming that the ancestors of the herbivorous giants of the Jurassic and Cretaceous Periods were small predators.

The end of the Triassic and the beginning of the Jurassic Periods were marked by the abundance of what have long been called "prosauropods," which were large herbivores found on every continent. This name has now been abandoned because prosauropods are not a monophyletic group (comprising the common ancestor and all the descendants of this common ancestor) but belong to several ancient branches of sauropodomorphs. Nevertheless, they have an unmistakable familial resemblance, with small, serrated leaf-shaped teeth, a small head, hind legs that were much more powerful than the front ones, and an elongated neck.

Among them, *Plateosaurus* was abundant in Europe during the Late Norian and Rhaetian Ages, the last two ages of the Triassic Period (around 214–204 Ma). It was also one of the first dinosaurs ever to be described back in 1834 by German paleontologist Hermann von Meyer (1801–1869).

The abundance of *Plateosaurus* skeletons in Germany has earned it the nickname *Schwäbischer Lindwurm* (Swabian lindworm, or dragon), in reference to local dragon legends, clearly inspired by ancient discoveries of bones! A discovery in Klagenfurt, Austria, offers an example: around 1335, when some young men found a fossilized rhinoceros skull in a cave, it was immediately believed to be a lindworm skull. The statue of this rhinoceros-headed lindworm is still on display today in the town square in Klagenfurt, a town whose coat of arms proudly displays this monster. *Plateosaurus* is also the hero of Franquin's comic strip *The Visitor from the Mesozoic*. If you haven't read it yet, it tells the story of a *Plateosaurus* egg discovered in Antarctica by Count Pacôme Hégésippe Adélard Ladislas de Champignac and of the adventures that ensued from bringing this egg back to the village of Champignac in France to hatch. While *Plateosaurus* isn't known for its intelligence, Franquin's plateosaur shows an appetite for play that's perhaps not so implausible.[9]

The first sauropods evolved from one of these primitive sauropodomorphs. While a few footprints discovered in Greenland[10] point to their appearance as early as the end of the Triassic Period, the oldest bony remains currently date from the end of the Early Jurassic Period, when sauropods were already large sizes (*Vulcanodon* from South Africa and *Tazoudasaurus* from Morocco measured 11 meters [36 feet] long).

APPENDIX

The eusauropods developed from the Middle Jurassic Period onward, including the strange mamenchisaurids with their immensely long necks. *Mamenchisaurus* had nineteen cervical vertebrae that measured 10 meters (32.8 feet) all together, for a total length of 22 meters (72.2 feet).

Neosauropods also arrived in the Middle Jurassic Period. During the Late Jurassic and Early Cretaceous, they could be found just about everywhere, with four emblematic families: diplodocids, brachiosaurids, camarasaurids, and dicraeosaurids. While they all shared a long neck, long tail, small head, and four legs, these families each showed characteristic variations from the bauplan of sauropods. Diplodocids had a long tail made up of seventy to eighty vertebrae; nostrils located on top of the skull; small, fine, cylinder-shaped teeth; and a tiny elongated head. Their forelegs were also much shorter than their hind legs, which means their rumps were up in the air. Diplodocids include *Diplodocus*, of course, and its American cousins *Barosaurus*, *Apatosaurus*, and *Brontosaurus*. Brachiosaurids, on the other hand, are easily recognized by their huge arms which are longer than their legs and their low buttocks. *Giraffatitan* (formerly part of the genus *Brachiosaurus*), from the Upper Jurassic of Tanzania, is the finest example of these five-story animals—its head reached a height of just over 13 meters (over 42 feet). In contrast, its cousin *Europasaurus* from Germany, which lived 155 million years ago, was 6 meters (19.7 feet) long at most and weighed 800 kilograms (1,764 pounds), a dwarf in the realm of giants. This dwarfism is probably related to *Europasaurus*'s insular habitat. It lived on one of the many islands of the European archipelago at the end of the Jurassic Period, and in this type of habitat with limited resources, evolution tends to reduce the size of large herbivores in order to conserve resources. Among the best-known examples of island dwarfism are the dwarf elephants and dwarf hippos that lived on Mediterranean islands tens of thousands of years ago. The same evolutionary phenomenon likely led to the dwarfism of *Europasaurus* and other European sauropods later on, such as the titanosaurs *Magyarosaurus* and *Lirainosaurus*.

Dicraeosaurids had short necks. *Brachytrachelopan* from the Upper Jurassic of Argentina had an especially short neck, and its body shape was only distantly reminiscent of sauropods. Its cousins *Amargasaurus* and *Bajadasaurus* had cervical vertebrae with immense neural spines that protruded from the skin of the neck.

All these sauropod groups disappeared during the Early Cretaceous Period, supplanted by the large branch of titanosaurs, which included the largest animals in Earth's history, as well as some of the smallest sauropods. Attempts to classify titanosaurs and neosauropods in general are still in their infancy; too many different hypotheses coexist for any certainties to emerge. The most that can be said is that brachiosaurids were the sister group to titanosaurs, and diplodocids were the sister group to dicraeosaurids. Beyond these conclusions, there are currently about as many hypotheses as there are researchers. Among the largest dinosaurs were *Patagotitan* and *Argentinosaurus*, two titanosaurs from South America that measured around 35 meters (115 feet) long and weighed over 70 metric tons. Being very large animals means their bones are ridiculously heavy and therefore extremely tiring for the average paleontologist to lug around. One sauropod femur discovered at Angeac in western France measured 2.2 meters (7.2 feet) in height and weighed over 100 kilograms (220 pounds). This is the kind of creature that can't be studied without prior weight-lifting training.

Thyreophorans

Thyreophorans include all stegosaurs and ankylosaurs. The oldest thyreophorans date from the beginning of the Jurassic Period, such as *Scutellosaurus*, which lived during the Sinemurian Age (199–190 Ma) in what is now Arizona. This little biped measured

1.2 meters (almost 4 feet) long and 50 centimeters (1.6 feet) at the hip, and weighed less than 10 kilograms (22 pounds), but it already possessed the main characteristic of its descendants: a body covered with several hundred osteoderms arranged in several rows. Its tail was unusually long, comprising almost two-thirds of its total body length, no doubt to maintain the balance of a body weighed down by its bony armor. *Scelidosaurus* was discovered in England (it lived during the Sinemurian to Pliensbachian Ages [199–182 Ma]), and with a length of 4 meters (about 13 feet) and weight of almost 300 kilograms (over 660 pounds), this quadruped would have already resembled an ankylosaur, which lived later.

The oldest stegosaur bone remains date from the Middle Jurassic Period. *Huayangosaurus*, discovered in China, was a small stegosaur measuring 4.5 meters (14.8 feet) long, with a double row of spines and plates along its neck, back, and tail that ended in two large pairs of spines forming the thagomizer, a characteristic of most stegosaurian dinosaurs. The name for this defensive spiny structure was invented by American cartoonist Gary Larson in his comic strip *The Far Side*, in a cartoon depicting prehistoric men attending a paleontology class. The teacher (also a prehistoric man, of course) presents the spines of a *Stegosaurus* tail to his students, explaining that this structure is called a "thagomizer" in memory of the late Thag Simmons (who was presumably its victim). As some paleontologists have a sense of humor, they adopted this name for this distinctive anatomical structure. At the end of the Jurassic Period, *Stegosaurus* in North America, *Miragaia* in Europe, and *Kentrosaurus* in Africa exemplified the apogee of the group, which seemed to disappear during the Early Cretaceous Period. Fossilized footprints discovered in 1995 from the basal Jurassic of the Dordogne department of France displayed all the characteristics expected of stegosaurian footprints and could therefore testify to the older age of the family.

Ankylosaurs are the other major group of thyreophorans. They are divided into two families: ankylosaurids and nodosaurids. The oldest ankylosaur is *Sarcolestes* from the Callovian Age of England. While they all looked like army tanks with their stocky osteoderm-covered bodies, Cretaceous ankylosaurids are most easily recognized by the bony club at the end of their tails. A century of illustrations depicting herbivorous *Ankylosaurus* has made us think about things from its perspective and how it used this club to strike tyrannosaurs in the legs to protect itself from being eaten. In any case, the oldest known wielders of this weapon are fairly recent, since the oldest known ankylosaurid, *Gastonia*, which lived around 135 million years ago, didn't have one. It was mainly in the second half of the Cretaceous Period (100–66 Ma) that this family of dinosaurs flourished in Asia and North America.

Nodosaurids didn't have their cousins' caudal club. This family includes one of the first three dinosaurs described in the 19th century, *Hylaeosaurus*, discovered by Gideon Mantell in 1832. Nodosaurids are known to have existed since the end of the Jurassic Period, including *Mymoorapelta* from what is now the United States, and to have developed until the end of the Cretaceous Period. A remarkable characteristic of nodosaurids was their powerful parascapular spine, a large pointed bone protruding from their shoulders. In *Borealopelta* from the Lower Cretaceous of Canada, the preservation of its intestinal contents has revealed its love of ferns.

Ornithopods

The first ornithopods are thought to have appeared in the Middle Jurassic Period. The group includes hadrosaurs (duck-billed dinosaurs) and their cousins, but its classification is only partially established. Several families, such as the small dryosaurids, Elasmaria from the southern continents, and rhabdodontids from Europe, have existed, but the phylogenetic relationships within Ornithopoda remain rather obscure. The famous *Iguanodon* is the cousin of hadrosaurs, the largest ornithopod branch. The most famous hadrosaurs belong to the hadrosaurid family: *Tsintaosaurus, Parasaurolophus,* and *Hypacrosaurus* are lambeosaurines (the crested hadrosaur subfamily), while *Maiasaura* and *Edmontosaurus* are saurolophines.

Marginocephalians

Marginocephalians are the family of the famous *Triceratops* and its many cousins, such as pachycephalosaurs and psittacosaurs. Marginocephalians and ornithopods are both within the clade Cerapoda, the sister group to thyreophorans. Pachycephalosaurs are easily recognized by the thick layer of bone covering their skulls, giving them the false air of high-brow intellectuals. They were mainly found in Asia and North America during the Late Cretaceous Period.

Psittacosaurs are a family containing a single genus of small dinosaurs from Asia with a powerful beak, known from numerous skeletons. Among the ceratopsians are various families, such as protoceratopsids, leptoceratopsids, and ceratopsids. Ceratopsidae includes most of the ceratopsian stars: *Triceratops, Centrosaurus, Pachyrhinosaurus,* and company.

ACKNOWLEDGMENTS

I'd like to thank Jessica Serra for suggesting, between two COVID-19 lockdowns, that I write a book "that takes you back in time and into the head of a dinosaur." It was an intriguing proposition, and one that I thoroughly enjoyed making a reality. A big thank-you to my first readers (Christel, Jessica, Olivia, and Joanna), who spared you, dear readers, from my worst jokes. Finally, for their involuntary but vital contribution to paleontology, I would like to thank all the physicists and engineers who theorized and developed X-ray microtomography, microspectroscopy, synchrotrons, and other toys that paleontologists have hijacked from their original purpose.

NOTES

Foreword

1. Konrad Lorenz, "Principe d'une finalité de l'ordre du monde," in *L'homme dans le fleuve du vivant* (Paris: Flammarion, 1981).

1. Dinosauria

1. Maureen A. O'Leary et al., "The placental mammal ancestor and the post-K-Pg radiation of placentals," *Science* 339, no. 6120 (2013): 662–667.
2. Adrian P. Hunt et al., "Vertebrate coprolites," *New Mexico Museum of Natural History & Science* 57 (2012); Jean Le Loeuff, "Vomi, caca, pipi, prout et crotte de nez—les mots de la paléocradologie," Le Dinoblog, February 7, 2013, https://www.dinosauria.org/blog/2013/02/07/vomi-caca-pipi-prout-et-crotte-de-nez-les-mots-de-la-paleocradologie/.
3. Marc E. H. Jones et al., "Integration of molecules and new fossils supports a Triassic origin for Lepidosauria (lizards, snakes, and tuatara)," *BMC Evolutionary Biology* 13, no. 208 (2013): 1–21.
4. Robert J. Asher, Nigel Bennett, and Thomas Lehmann, "The new framework for understanding placental mammal evolution," *BioEssays* 31, no. 8 (2009): 853–864.

2. Dinosaur Meninges

1. Lawrence M. Witmer et al., "Using CT to peer into the past: 3D visualization of the brain and ear regions of birds, crocodiles, and nonavian dinosaurs," in *Anatomical Imaging: Towards a New Morphology*, ed. Hideki Endo and Roland Frey (Tokyo: Springer, 2008), 67–87.
2. J. W. Hulke, "Note on a large reptilian skull from Brooke, Isle of Wight, probably dinosaurian, and referable to the genus *Iguanodon*," *Quarterly Journal of the Geological Society of London* 27, no. 1–2 (1871): 199–206.

3. Simon Wills, "John Whitaker Hulke, surgeon and palaeontologist," in *A History of Geology and Medicine*, ed. C. J. Duffin, R. T. J. Moody, and C. Gardner-Thorpe (London: Geological Society of London, 2013), 375.
4. Othniel Charles Marsh, "Principal characters of American Jurassic dinosaurs, part IV. Spinal cord, pelvis, and limbs of *Stegosaurus*," *American Journal of Science* 21, no. 122 (1881): 167–170.
5. Othniel Charles Marsh, "The dinosaurs of North America," *Sixteenth Annual Report of the U.S. Geological Survey*, part I (1896): 135–244 and plates ii–lxxxv.
6. Marcellin Boule, "Les grands animaux fossiles de l'Amérique," *Association française pour l'avancement des sciences: conférences de Paris, compte-rendu de la 20e session* (1891): 18–38.
7. Marcellin Boule, "Les créatures géantes d'autrefois," *Revue générale des sciences pures et appliquées* 13 (1902): 903–915.
8. Richard Swann Lull, "On the functions of the 'sacral brain' in dinosaurs," *American Journal of Science* 44, no. 264 (1917): 471–477.
9. Lowell Dingus and Mark Norell, *Barnum Brown: The Man Who Discovered Tyrannosaurus Rex* (Berkeley: University of California Press, 2010); Jean Le Loeuff, *T. rex superstar* (Paris: Belin, 2016).
10. Barnum Brown, "*Anchiceratops*, a new genus of horned dinosaurs from the Edmonton Cretaceous of Alberta; with discussion of the origin of the ceratopsian crest and the brain casts of *Anchiceratops* and *Trachodon*," *Bulletin of the AMNH* 33 (1914): 539–548.
11. Richard Carrington, *A Guide to Earth History* (London: Chatto and Windus, 1956).
12. Emily A. Buchholtz and Ernst-August Seyfarth, "The gospel of the fossil brain: Tilly Edinger and the science of paleoneurology," *Brain Research Bulletin* 48, no. 4 (1999): 351–361.
13. Tilly Edinger, "Anthropocentric misconceptions in paleoneurology," *Proceedings of the Rudolf Virchow Medical Society in the City of New York* 19 (1960): 56–107.
14. Harry J. Jerison, *Evolution of the Brain and Intelligence* (New York: Academic Press, 1973).

15. Gerhard Roth and Ursula Dicke, "Evolution of the brain and intelligence," *Trends in Cognitive Sciences* 9, no. 5 (2005): 250–257.
16. Daniel Jirak and Jiri Janacek, "Volume of the crocodilian brain and endocast during ontogeny," *PLoS ONE* 12, no. 6 (2017): e0178491.
17. Roger B. J. Benson et al., "Rates of dinosaur body mass evolution indicate 170 million years of sustained ecological innovation on the avian stem lineage," *PLoS Biology* 12, no. 5 (2014): e1001853.
18. Ashley Morhardt, "Gross anatomical brain region approximation (GABRA): assessing brain size, structure, and evolution in extinct archosaurs," (PhD diss., Ohio University, 2016).
19. Suzana Herculano-Houzel, "Numbers of neurons as biological correlates of cognitive capability," *Current Opinion in Behavioral Sciences* 16 (2017): 1–7.
20. Martin D. Brasier et al., "Remarkable preservation of brain tissues in an Early Cretaceous iguanodontian dinosaur," *Geological Society of London* 448 (2017): 383–398.
21. Dale A. Russell and R. Séguin, "Reconstruction of the small Cretaceous theropod *Stenonychosaurus inequalis* and a hypothetical dinosauroid," *Syllogeus* 37 (1982): 1–43.
22. D. J. Varricchio, J. D. Hogan, and W. J. Freimuth, "Revisiting Russell's troodontid: autecology, physiology, and speculative tool use," *Canadian Journal of Earth Sciences* 58, no. 9 (2021): 796–811.
23. Gavin R. Hunt, "Manufacture and use of hook-tools by New Caledonian crows," *Nature* 379 (1996): 249–251.
24. Jean-Luc Marcastel, *Tellucidar*, vol. 1 (Paris: Scrineo, 2016).
25. Dominique Delpiroux, *Les doigts du Diable* (Paris: L'Écailler, 2011).
26. Gavin A. Schmidt and Adam Frank, "The Silurian hypothesis: would it be possible to detect an industrial civilization in the geological record?," *International Journal of Astrobiology* 18, no. 2 (2019): 142–150.
27. Schmidt and Frank, "The Silurian hypothesis."
28. Robert Silverberg, "Our lady of the sauropods," *Omni*, September 1980.
29. Tilly Edinger, "The pituitary body in giant animals fossil and living: a survey and a suggestion," *Quarterly Review of Biology* 17, no. 1 (1942): 31–45.

30. Grant Richard Hurlburt, "Relative brain size in recent and fossil amniotes: determination and interpretation" (PhD diss., University of Toronto, 1996).

3. In Search of Lost Senses

1. Benoît Grison, *Les portes de la perception animale* (Lonay, Delachaux et Niestlé, 2021).
2. Darla K. Zelenitsky, François Therrien, and Yoshitsugu Kobayashi, "Olfactory acuity in theropods: palaeobiological and evolutionary implications," *Proceedings of the Royal Society B* 276, no. 1657 (2009): 667–673; Darla K. Zelenitsky et al., "Evolution of olfaction in non-avian theropod dinosaurs and birds," *Proceedings of the Royal Society B* 278, no. 1725 (2011): 3625–3634.
3. Graham M. Hughes and John A. Finarelli, "Olfactory receptor repertoire size in dinosaurs," *Proceedings of the Royal Society B* 286, no. 1904 (2019).
4. Rina Sakagami and Soichiro Kawabe, "Endocranial anatomy of the ceratopsid dinosaur *Triceratops* and interpretations of sensory and motor function," *PeerJ* 8 (2020): e9888.
5. Tetsuto Miyashita et al., "The internal cranial morphology of an armoured dinosaur *Euoplocephalus* corroborated by X-ray computed tomographic reconstruction," *Journal of Anatomy* 219, no. 6 (2011): 661–675.
6. Victoria M. Arbour and Jordan C. Mallon, "Unusual cranial and postcranial anatomy in the archetypal ankylosaur *Ankylosaurus magniventris*," *FACETS* 2, no. 2 (2017): 764–794.
7. Paul J. Weldon and James W. Wheeler, "The chemistry of crocodilian skin glands," in *Crocodilian Biology and Evolution*, ed. Gordon C. Grigg, Frank Seebacher, and Craig E. Franklin (Chipping Norton: Surrey Beatty & Sons, 2001), 286–296.
8. Jakob Vinther, Robert Nicholls, and Diane A. Kelly, "A cloacal opening in a non-avian dinosaur," *Current Biology* 31, no. 4 (2021): R182–R183.
9. David M. Wilkinson, Euan G. Nisbet, and Graeme D. Ruxton, "Could methane produced by sauropod dinosaurs have helped drive Mesozoic climate warmth?," *Current Biology* 22, no. 9 (2012): R292–R293.
10. Lawrence M. Witmer and Ryan C. Ridgely, "New insights into the brain, braincase, and ear region of tyrannosaurs (Dinosauria, Theropoda), with

implications for sensory organization and behavior," *Anatomical Record* 292, no. 9 (2009): 1266–1296.

11. Fabien Knoll et al., "A new titanosaurian braincase from the Cretaceous 'Lo Hueco' locality in Spain sheds light on neuroanatomical evolution within Titanosauria," *PLoS ONE* 10, no. 10 (2015): e0138233.

12. Lars Schmitz and Ryosuke Motani, "Nocturnality in dinosaurs inferred from scleral ring and orbit morphology," *Science* 332, no. 6030 (2011): 705–708.

13. J. Logan King et al., "The endocranium and trophic ecology of *Velociraptor mongoliensis*," *Journal of Anatomy* 237, no. 5 (2020): 861–869.

14. Jonah N. Choiniere et al., "Evolution of vision and hearing modalities in theropod dinosaurs," *Science* 372, no. 6542 (2021): 610–613.

15. Nicolas Nagloo et al., "Spatial resolving power and spectral sensitivity of the saltwater crocodile, *Crocodylus porosus*, and the freshwater crocodile, *Crocodylus johnstoni*," *Journal of Experimental Biology* 219, pt. 9 (2016): 1394–1404.

16. Ryan M. Carney, Helmut Tischlinger, and Matthew D. Shawkey, "Evidence corroborates identity of isolated fossil feather as a wing covert of *Archaeopteryx*," *Scientific Reports* 10, no. 15593 (2020).

17. Fucheng Zhang et al., "Fossilized melanosomes and the colour of Cretaceous dinosaurs and birds," *Nature* 463 (2010): 1075–1078.

18. Caleb M. Brown et al., "An exceptionally preserved three-dimensional armored dinosaur reveals insights into coloration and Cretaceous predator-prey dynamics," *Current Biology* 27, no. 16 (2017): 2514–2521.

19. Maria E. McNamara et al., "Reconstructing carotenoid-based and structural coloration in fossil skin," *Current Biology* 26, no. 8 (2016): 1075–1082.

20. Guy Costes and Joseph Altairac, *Rétrofictions: Encyclopédie de la conjecture romanesque rationnelle francophone, de Rabelais à Barjavel, 1532–1951* (Paris: Les Belles Lettres, 2018).

21. Philippe Brux and Georges Kouroussis, "Les répulsifs électroniques," *Nuisibles et parasites information* 73 (2012): 23–25.

22. Otto Gleich, Robert J. Dooling, and Geoffrey A. Manley, "Audiogram, body mass, and basilar papilla length: correlations in birds and predictions for extinct archosaurs," *Naturwissenschaften* 92 (2005): 595–598.

23. Jonah N. Choiniere et al., "Evolution of vision and hearing modalities in theropod dinosaurs," *Science* 372, no. 6542 (2021): 610–613.
24. Camille Flammarion, *Le monde avant la création de l'homme* (Paris: Emile Levy, 1886).
25. Phil Senter, "Voices of the past: a review of Paleozoic and Mesozoic animal sounds," *Historical Biology* 20, no. 4 (2008): 255–287.
26. A. L. Vergne, M. B. Pritz, and N. Mathevon, "Acoustic communication in crocodilians: from behaviour to brain," *Biological Reviews of the Cambridge Philosophical Society* 84, no. 3 (2009): 391–411.
27. Tobias Riede et al., "Coos, booms, and hoots: the evolution of closed-mouth vocal behavior in birds," *Evolution* 70, no. 8 (2016): 1734–1746.
28. David B. Weishampel, "Acoustic analyses of potential vocalization in lambeosaurine dinosaurs (Reptilia: Ornithischia)," *Paleobiology* 7, no. 2 (1981): 252–261.
29. David C. Evans, Ryan Ridgely, and Lawrence M. Witmer, "Endocranial anatomy of lambeosaurine hadrosaurids (Dinosauria: Ornithischia): a sensorineural perspective on cranial crest function," *Anatomical Record* 292, no. 9 (2009): 1315–1337.
30. Serge Gainsbourg, "Sois belle et tais-toi," 1960.
31. Nicolas Di-Poï and Michel C. Milinkovitch, "Crocodylians evolved scattered multi-sensory micro-organs," *EvoDevo* 4 (2013): 19.
32. Thomas D. Carr et al., "A new tyrannosaur with evidence for anagenesis and crocodile-like facial sensory system," *Scientific Reports* 7 (2017): 44942.
33. Phil R. Bell and Christophe Hendrickx, "Crocodile-like sensory scales in a Late Jurassic theropod dinosaur," *Current Biology* 30, no. 19 (2020): R1068–R1070.
34. Les Hearn and Amanda C. de C. Williams, "Pain in dinosaurs: what is the evidence?," *Philosophical Transactions of the Royal Society B* 374, no. 1785 (2019).
35. Darren H. Tanke and Bruce M. Rothschild, "Paleopathology in Late Cretaceous Hadrosauridae from Alberta," in *Hadrosaurs*, ed. David A. Eberth and David C. Evans (Bloomington, IN: Indiana University Press, 2014), 540–571.

36. Richard T. McCrea et al., "Vertebrate ichnopathology: pathologies inferred from dinosaur tracks and trackways from the Mesozoic," *Ichnos* 22, no. 3–4 (2015): 235–260.
37. Seper Ekhtiari et al., "First case of osteosarcoma in a dinosaur: a multimodal diagnosis," *Lancet Oncology* 21, no. 8 (2020): P1021–P1022.
38. Ewan D. S. Wolff et al., "Common avian infection plagued the tyrant dinosaurs," *PLoS ONE* 4, no. 9 (2009): e7288.
39. K. S. Brink, "Description of new tooth pathologies in *Tyrannosaurus rex*," *Journal of Vertebrate Paleontology, Program and Abstracts* (2020): 84–85.
40. Bruce M. Rothschild and Robert Depalma, "Skin pathology in the Cretaceous: evidence for probable failed predation in a dinosaur," *Cretaceous Research* 42 (2013): 44–47.
41. Paul-Antoine Libourel and Baptiste Barrillot, "Is there REM sleep in reptiles? A key question, but still unanswered," *Current Opinion in Physiology* 15 (2020): 134–142.
42. Nadine Gravett et al., "Inactivity/sleep in two wild free-roaming African elephant matriarchs—does large body size make elephants the shortest mammalian sleepers?," *PLoS ONE* 12, no. 3 (2017): e0171903.
43. Xing Xu and Mark A. Norell, "A new troodontid dinosaur from China with avian-like sleeping posture," *Nature* 431 (2004): 838–841.
44. Romain Amiot et al., "Oxygen isotopes from biogenic apatites suggest widespread endothermy in Cretaceous dinosaurs," *Earth and Planetary Science Letters* 246, no. 1–2 (2006): 41–54.
45. Robin R. Dawson et al., "Eggshell geochemistry reveals ancestral metabolic thermoregulation in Dinosauria," *Science Advances* 6, no. 7 (2020).

4. Mesozoic Sociology

1. J. Sean Doody, Gordon M. Burghardt, and Vladimir Dinets, "Breaking the social–non-social dichotomy: a role for reptiles in vertebrate social behavior research?," *Ethology* 119, no. 2 (2013): 95–103.
2. E. Hennig, "*Kentrosaurus aethiopicus* der Stegosauride des Tendaguru," *Sitzungsberichte der Gesellschaft Naturforschender Freunde zu Berlin* (1915): 219–247.

3. Victoria M. Arbour and Jordan C. Mallon, "Unusual cranial and postcranial anatomy in the archetypal ankylosaur *Ankylosaurus magniventris*," *FACETS* 2, no. 2 (2017): 764–794.
4. Jean Le Loeuff, *T. rex superstar* (Paris: Belin, 2016).
5. Roland T. Bird, *Bones for Barnum Brown: Adventures of a Dinosaur Hunter* (Fort Worth, TX: Texas Christian University Press, 1985).
6. David A. Eberth, D. B. Brinkman, and V. Barkas, "A centrosaurine megabonebed from the Upper Cretaceous of southern Alberta: implications for behavior and death events," in *New Perspectives on Horned Dinosaurs*, ed. Michael J. Ryan, Brenda J. Chinnery-Allgeier, and David A. Eberth (Bloomington, IN: Indiana University Press, 2010): 495–508.
7. Joshua C. Mathews et al., "The first triceratops bonebed and its implications for gregarious behavior," *Journal of Vertebrate Paleontology* 29, no. 1 (2009): 286–290.
8. Shay Gueron and Simon A. Levin, "Self-organization of front patterns in large wildebeest herds," *Journal of Theoretical Biology* 165, no. 4 (1993): 541–552.
9. Yaoming Hu et al., "Large Mesozoic mammals fed on young dinosaurs," *Nature* 433 (2005): 149–152.
10. Zhao Qi, Paul M. Barrett, and David A. Eberth, "Social behaviour and mass mortality in the basal ceratopsian dinosaur *Psittacosaurus* (early Cretaceous, People's Republic of China)," *Palaeontology* 50, no. 5 (2007): 1023–1029.; Brandon P. Hedrick et al., "The osteology and taphonomy of a *Psittacosaurus* bonebed assemblage of the Yixian Formation (Lower Cretaceous), Liaoning, China," *Cretaceous Research* 51 (2014): 321–340.
11. Adrienne Mayor, *The First Fossil Hunters: Paleontology in Greek and Roman Times* (Princeton, NJ: Princeton University Press, 2000).
12. David W. E. Hone et al., "A new mass mortality of juvenile *Protoceratops* and size-segregated aggregation behaviour in juvenile non-avian dinosaurs," *PLoS ONE* 9, no. 11 (2014): e113306.
13. David J. Varricchio et al., "Mud-trapped herd captures evidence of distinctive dinosaur sociality," *Acta Palaeontologica Polonica* 53, no. 4 (2008): 567–578.

14. Robert A. DePalma II et al., "Physical evidence of predatory behavior in *Tyrannosaurs rex*," *PNAS* 110, no. 31 (2013): 12560–12564.
15. Nicholas R. Longrich et al., "Cannibalism in *Tyrannosaurus rex*," *PLoS ONE* 5, no 10 (2010): e13419.
16. Raymond R. Rogers, David W. Krause, and Kristina Curry Rogers, "Cannibalism in the Madagascan dinosaur *Majungatholus atopus*," *Nature* 422 (2003): 515–518.
17. Jean Le Loeuff et al., "The first dinosaur footprints from the Khok Kruat Formation (Aptian of northeastern Thailand)," *Mahasarakham University Journal* 22 (2003): 83–91.
18. Brian T. Roach and Daniel L. Brinkman, "A reevaluation of cooperative pack hunting and gregariousness in *Deinonychus antirrhopus* and other nonavian theropod dinosaurs," *Bulletin of the Peabody Museum of Natural History* 48, no. 1 (2007): 103–138.
19. J. A. Frederickson, M. H. Engel, and R. L. Cifelli, "Ontogenetic dietary shifts in *Deinonychus antirrhopus* (Theropoda; Dromaeosauridae): insights into the ecology and social behavior of raptorial dinosaurs through stable isotope analysis," *Palaeogeography, Palaeoclimatology, Palaeoecology* 552 (2020): 109780.
20. Zofia Kielan-Jaworowska and Rinchen Barsbold, "Narrative of the Polish-Mongolian palaeontological expeditions 1967–1971," *Palaeontologia Polonica* 27 (1972): 5–136.
21. Kenneth Carpenter, "Evidence of predatory behavior by carnivorous dinosaurs," *Gaia* 15 (1998): 135–144.
22. Cameron C. Pahl and Luis A. Ruedas, "Carnosaurs as apex scavengers: agent-based simulations reveal possible vulture analogues in late Jurassic dinosaurs," *Ecological Modelling* 458 (2021): 109706.
23. Jakob Vinther et al., "3D camouflage in an ornithischian dinosaur," *Current Biology* 26, no. 18 (2016): 2456–2462.
24. Jean-François Bouvet, *La stratégie du caméléon* (Paris: Seuil, 2000).
25. Caleb M. Brown et al., "An exceptionally preserved three-dimensional armored dinosaur reveals insights into coloration and cretaceous predator-prey dynamics," *Current Biology* 27, no. 16 (2017): 2514–2521.

26. Victoria Megan Arbour, "Estimating impact forces of tail club strikes by ankylosaurid dinosaurs," *PLoS ONE* 4, no. 8 (2009): e6738.
27. Karen Chin et al., "A king-sized theropod coprolite," *Nature* 393 (1998): 680–682.
28. Andrew A. Farke, Ewan D. S. Wolff, and Darren H. Tanke, "Evidence of combat in *Triceratops*," *PLoS ONE* 4, no 1 (2009): e4252.
29. David F. Terrill, Charles M. Henderson, and Jason S. Anderson, "New application of strontium isotopes reveals evidence of limited migratory behaviour in Late Cretaceous hadrosaurs," *Biology Letters* 16, no. 3 (2020).
30. Anthony R. Fiorillo, Stephen T. Hasiotis, and Yoshitsugu Kobayashi, "Herd structure in Late Cretaceous polar dinosaurs: a remarkable new dinosaur tracksite, Denali National Park, Alaska, USA," *Geology* 42, no. 8 (2014): 719–722.
31. Walter Scott Persons IV and Philip J. Currie, "Duckbills on the run: the cursorial abilities of hadrosaurs and implications for tyrannosaur-avoidance strategies," in *Hadrosaurs*, ed. David A. Eberth and David C. Evans (Bloomington, IN: Indiana University Press, 2014): 449–458.
32. Julia J. Day et al., "Dinosaur locomotion from a new trackway," *Nature* 415 (2002): 494–495.
33. James O. Farlow, "Estimates of dinosaur speeds from a new trackway site in Texas," *Nature* 294 (1981): 747–748.
34. Martin G. Lockley et al., "A preliminary report of an Early Jurassic *Eubrontes*-dominated tracksite in the Navajo Sandstone Formation at the mail station dinosaur tracksite, San Juan County, Utah," *New Mexico Museum of Natural History and Science Bulletin* 82 (2021): 195–208.
35. R. McNeil Alexander et al., "Mechanics of running of the ostrich (*Struthio camelus*)," *Journal of Zoology* 187, no. 2 (1979): 169–178.
36. Karen Chin, Rodney M. Feldmann, and Jessica N. Tashman, "Consumption of crustaceans by megaherbivorous dinosaurs: dietary flexibility and dinosaur life history strategies," *Scientific Reports* 7 (2017): 11163.
37. Nicolas Mengden, *L'Aventure en images, no. 3: L'île de l'épouvante* (Paris: Les Éditions modernes, 1938).
38. Caleb M. Brown et al., "Dietary palaeoecology of an Early Cretaceous armoured dinosaur (Ornithischia; Nodosauridae) based on floral

analysis of stomach contents," *Royal Society Open Science* 7, no. 6 (2020): 200305.
39. Andrew C. Scott et al., *Fire on Earth: An Introduction* (Hoboken, NJ: Wiley-Blackwell, 2014).
40. Robert V. Hill et al., "A complex hyobranchial apparatus in a Cretaceous dinosaur and the antiquity of avian paraglossalia," *Zoological Journal of the Linnean Society* 175, no. 4 (2015): 892–909.
41. William J. Freimuth et al., "Mammal-bearing gastric pellets potentially attributable to *Troodon formosus* at the Cretaceous Egg Mountain locality, Two Medicine Formation, Montana, USA," *Palaeontology* 64, no. 5 (2021): 699–725.

5. Banter Between Lovers

1. J. C. Mallon, "Recognizing sexual dimorphism in the fossil record: lessons from nonavian dinosaurs," *Paleobiology* 43, no. 3 (2017): 495–507.
2. L. Sprague de Camp, "A gun for dinosaur," *Galaxy Science Fiction*, March 1956, 7.
3. Joseph E. Peterson and Christopher P. Vittore, "Cranial pathologies in a specimen of *Pachycephalosaurus*," *PLoS ONE* 7, no. 4 (2012): e36227.
4. D. H. Tanke and B. M. Rothschild, "Paleopathologies in Albertan ceratopsids and their behavioral significance," in *New Perspectives on Horned Dinosaurs*, ed. Michael J. Ryan, Brenda J. Chinnery-Allgeier, and David A. Eberth (Bloomington, IN: Indiana University Press, 2010): 355–384.
5. Caleb M. Brown, Philip J. Currie, and François Therrien, "Intraspecific facial bite marks in tyrannosaurids provide insight into sexual maturity and evolution of bird-like intersexual display," *Paleobiology* 48, no. 1 (2022): 12–43.
6. Katja Waskow and P. Martin Sander, "Growth record and histological variation in the dorsal ribs of *Camarasaurus* sp. (Sauropoda)," *Journal of Vertebrate Paleontology* 34, no. 4 (2014): 852–869.
7. Gregory M. Erickson et al., "Gigantism and comparative life-history parameters of tyrannosaurid dinosaurs," *Nature* 430 (2004): 772–775.
8. Eva Maria Griebeler, Nicole Klein, and P. Martin Sander, "Aging, maturation and growth of sauropodomorph dinosaurs as deduced from

growth curves using long bone histological data: an assessment of methodological constraints and solutions," *PLoS ONE* 8, no. 6 (2013): e67012.

9. Francis B. Nopcsa, "Remarks on the supposed clavicle of the sauropodous dinosaur *Diplodocus*," *Proceedings of the Zoological Society of London* 75, no. 3 (1905): 289–294.

10. Thomas Ziegler and Sven Olbort, "Genital structures and sex identification in crocodiles," *Crocodile Specialist Group Newsletter* (2007).

11. Jakob Vinther, Robert Nicholls, and Diane A. Kelly, "A cloacal opening in a non-avian dinosaur," *Current Biology* 31, no. 4 (2021): R182–R183.

12. Timothy E. Isles, "The socio-sexual behaviour of extant archosaurs: implications for understanding dinosaur behaviour," *Historical Biology* 21, no. 3–4 (2009): 139–214.

13. Filippo Bertozzo, "A tail tale: injuries in caudal neural spines of Hadrosauridae revealed by an extensive paleopathological revision of Ornithopoda," *Journal of Vertebrate Paleontology* (2020): 75.

14. "Dinosaurs cavort in film for Doyle," *New York Times*, June 3, 1922.

15. Fernand Mysor, *Les semeurs d'épouvante, Roman des Temps Jurassiques* (Paris: Grasset, 1923).

16. Grom, T-rex *mon amour* (Grom Books, 2020).

17. Jean Le Loeuff, "L'Abbé Pouech et les dinosaures du Plantaurel," in *Actes du Colloque Jean-Jacques Pouech* (Pamiers, France: Imprimerie Polito & Fils, 1993), 23–30.; É. Buffetaut and Jean Le Loeuff, "The discovery of dinosaur eggshells in nineteenth-century France," in *Dinosaur Eggs and Babies*, ed. Kenneth Carpenter, Karl F. Hirsch, and John R. Horner (Cambridge, UK: Cambridge University Press, 1996), 31–34.

18. Bernat Vila et al., "3-D modelling of megaloolithid clutches: insights about nest construction and dinosaur behaviour," *PLoS ONE* 5, no. 5 (2010): e10362.

19. "Sauce boat," Wikipedia, last modified February 11, 2023, https://en.wikipedia.org/wiki/Sauce_boat.

20. Robert Duncan Milne, "The Iguanodon's Egg," *The Argonaut*, 1882.

21. Mark A. Norell et al., "The first dinosaur egg was soft," *Nature* 583 (2020): 406–410.

22. Jasmina Wiemann et al., "Dinosaur origin of egg color: oviraptors laid blue-green eggs," *PeerJ* 5 (2017): e3706; Jasmina Wiemann, Tzu-Ruei Yang, and Mark A. Norell, "Dinosaur egg colour had a single evolutionary origin," *Nature* 563, no. 7732 (2018): 555–558.
23. David J. Varricchio, John R. Horner, and Frankie D. Jackson, "Embryos and eggs for the Cretaceous theropod dinosaur *Troodon formosus*," *Journal of Vertebrate Paleontology* 22, no. 3 (2002): 564–576.
24. Gregory M. Erickson et al., "Dinosaur incubation periods directly determined from growth-line counts in embryonic teeth show reptilian-grade development," *PNAS* 114, no. 3 (2017): 540–545.
25. David J. Varricchio, Martin Kundrát, and Jason Hogan, "An intermediate incubation period and primitive brooding in a theropod dinosaur," *Scientific Reports* 8 (2018): 12454.
26. Amélie L. Vergne et al., "Parent-offspring communication in the Nile crocodile *Crocodylus niloticus*: do newborns' calls show an individual signature?," *Naturwissenschaften* 94 (2007): 49–54.
27. T-R Yang et al., "Hatching asynchrony in oviraptorid dinosaurs sheds light on their unique nesting biology," *Integrative Organismal Biology* 1, no. 1 (2019): obz030; Shundong Bi et al., "An oviraptorid preserved atop an embryo-bearing egg clutch sheds light on the reproductive biology of non-avialan theropod dinosaurs," *Science Bulletin* 66, no. 9 (2021): 947–954.
28. Gordon M. Burghardt, *The Genesis of Animal Play: Testing the Limits* (Cambridge, MA: MIT Press, 2005).
29. Vladimir Dinets, "Play behavior in crocodilians," *Animal Behavior and Cognition* 2, no. 1 (2015): 49–55.
30. Bruce M. Rothschild, "Unexpected behavior in the Cretaceous: tooth-marked bones attributable to tyrannosaur play," *Ethology Ecology & Evolution* 27, no. 3 (2014): 325–334.
31. Joseph Peterson et al., "Face biting on a juvenile tyrannosaurid and behavioral implications," *Palaios* 24 (2009): 780–784.

Epilogue

1. Tom van der Valk et al., "Million-year-old DNA sheds light on the genomic history of mammoths," *Nature* 591 (2021): 265–269.

Appendix

1. Ricardo N. Martinez et al., "A basal dinosaur from the dawn of the dinosaur era in southwestern Pangaea," *Science* 331, no. 6014 (2011): 206–210.
2. Eric Buffetaut et al., "The earliest known sauropod dinosaur," *Nature* 407 (2000): 72–74; Andrew Racey and Jeffery G. S. Goodall, "Palynology and stratigraphy of the Mesozoic Khorat Group red bed sequences from Thailand," *Geological Society of London* 315 (2009): 69–83.
3. Sterling J. Nesbitt et al., "The oldest dinosaur? A Middle Triassic dinosauriform from Tanzania," *Biology Letters* 9, no. 1 (2013).
4. Martin Qvarnström et al., "Beetle-bearing coprolites possibly reveal the diet of a Late Triassic dinosauriform," *Royal Society Open Science* 6, no. 3 (2019).
5. Novella L. Razzolini et al., "Ichnological evidence of megalosaurid dinosaurs crossing Middle Jurassic tidal flats," *Scientific Reports* 6 (2016): 31494.
6. Jacques-Amand Eudes-Deslongchamps, "Mémoire sur le *Poekilopleuron bucklandii*, grand saurien fossile, intermédiaire entre les crocodiles et les lézards, découvert dans les carrières de la Maladrerie, près Caen, au mois de juillet 1835," *Mémoires de la Société Linnéenne de Normandie* 6 (1838): 1–114.
7. Yuong-Nam Lee et al., "Resolving the long-standing enigmas of a giant ornithomimosaur *Deinocheirus mirificus*," *Nature* 515 (2014): 257–260; Jean Le Loeuff, "*Deinocheirus*, l'antimanchot qui a retrouvé son corps," Le Dinoblog, November 24, 2014, https://www.dinosauria.org/blog/2014/11/24/deinocheirus-lantimanchot-qui-a-retrouve-son-corps/.
8. Mike Cummings, "Yale's legacy in 'Jurassic World,'" *Yale News*, June 18, 2015, https://news.yale.edu/2015/06/18/yale-s-legacy-jurassic-world.
9. Franquin, *Le voyageur du Mésozoïque* (Paris: Dupuis, 1960).
10. Jens N. Lallensack et al., "Sauropodomorph dinosaur trackways from the Fleming Fjord Formation of East Greenland: evidence for Late Triassic sauropods," *Acta Palaeontologica Polonica* 62, no. 4 (2017): 833–843.